Hermann Haupt

General Theory of Bridge Construction

Hermann Haupt

General Theory of Bridge Construction

ISBN/EAN: 9783337215644

Printed in Europe, USA, Canada, Australia, Japan

Cover: Foto ©berggeist007 / pixelio.de

More available books at **www.hansebooks.com**

GENERAL THEORY

OF

BRIDGE CONSTRUCTION:

CONTAINING

DEMONSTRATIONS OF THE PRINCIPLES OF THE ART
AND THEIR APPLICATION TO PRACTICE;

FURNISHING THE MEANS OF

CALCULATING THE STRAINS UPON THE CHORDS, TIES, BRACES,
COUNTER-BRACES, AND OTHER PARTS OF A BRIDGE
OR FRAME OF ANY DESCRIPTION.

WITH PRACTICAL ILLUSTRATIONS.

BY

HERMAN HAUPT, A.M.,
CIVIL ENGINEER.

NEW YORK:
D. APPLETON AND COMPANY,
1, 3, AND 5 BOND STREET.
1886.

ENTERED, according to Act of Congress, in the year 1851, by
D. APPLETON & CO.,
In the Clerk's Office of the District Court of the United States for the Southern District of New York.

PREFACE.

From the great importance of the art of bridge construction, it might be supposed that its principles would be familiarly understood by all whose occupations require an acquaintance with it. If, however, any work exists, containing an exposition of a theory sufficient to account generally for the various phenomena observed in the mutual action of the parts of trussed combinations of wood or metals, the author has neither seen or heard of it. When his attention was first directed to the subject of properly proportioning the parts of bridges, by being called upon in the discharge of professional duties to superintend their construction, he found it impossible to procure satisfactory information, either from engineers and builders, or from books. In fact, if he may be permitted to judge from the contradictory opinions that are even yet entertained, and from the errors frequently committed by practical men in the construction of bridges, he would be inclined to infer, that of few arts of equal practical importance, are the principles so little understood.

The best works on the subject of construction, that have fallen into the hands of the writer, contain but little that will furnish the means of calculating the strains upon the timbers of a bridge truss, or of determining their relative sizes; and they do not furnish information as to what constitute the elements of a framed truss, or the most advantageous disposition of these elements to attain the maximum strength and stiffness with a given quantity of material.

In the following pages, the object has been, not so much to detail particular plans, as to establish general principles.

An attempt has been made to explain the mode of action of the parts of structures, and their mutual influence when combined; to point out the ways by which the strains can be estimated, and the relative sizes of the timbers accurately determined; new combinations of the elements in the construction of bridge trusses have been suggested, the defects of many plans in general use pointed out, and several simple means proposed for remedying these defects and adding to the strength of structures.

Unable to procure satisfactory information in any other way, the writer, in the spring of 1840, commenced a course of experiments on models, and examined many existing structures, for which his occupation at that time as a civil engineer in the service of the **State of** Pennsylvania, afforded great facilities.

This investigation was commenced without the most remote design of ever giving the **result publicity, but solely for the purpose of enabling him to proceed** intelligently in **the discharge of professional duties.** The facts elicited, **led him** to conclude that many important structures, and some **of** recent erection, exhibited defects **which,** although serious, had escaped the observation of merely practical **builders;** these will be pointed out in their proper places, **and the** writer will be **amply** rewarded for the labor of preparing this little treatise, if it shall prove the means of adding in the smallest degree to the **stock of** information already possessed on **the important subject of bridge construction.**

The treatise on bridge construction has been prefaced by a few **pages on the resistance** of solids, because a **knowledge** of this subject lies at the foundation of **the art of construction.** The mode **of investigation is peculiar, and is** believed to be more simple than those usually employed.

This part of the subject will not be interesting to those who are unacquainted with the application of mathematics to mechanics, but those who are, will perhaps be pleased **with** the simplicity of the solutions, and the novelty of representing strains by geometrical solids, deflections by parabolic areas, and **the variable** pressures at different points of beams by the corresponding ordinates of plane **curves. The portion** which refers **to** the principles of construction, and their application to practice, will be readily understood by the general reader. Demonstrations of propositions required in the solution of problems, and not found in any work accessible to the author, have been given in notes.

It did not form part of the original design to include the subject of the construction of stone arches, but it was thought that the work would be rendered more useful, complete, and generally acceptable, if a few pages were added containing a simple exposition of the principles of this important art.

Mathematicians have apparently exhausted their ingenuity in devising modes of distributing the weights so as to produce an equilibrated curve of suitable form for the intrados of an arch; but many of their speculations are far more curious than useful, whilst practical men have been disposed to reject the principle of equilibration as inapplicable to constructions. It was a long time before the fortunate discovery was made that the intrados might be of any form most pleasing to the eye, and that the conditions of equilibrium could, in general, be satisfied by making the joints of the voussoirs perpendicular to the line of direction of the pressures; a fact so simple and obvious, that there is reason for surprise that it was not suggested to the first mind in which originated the idea of an arch of equilibrium. This principle is important in its practical results, and an admirable application is made of it, by John Seaward, a British Engineer, in a work containing a proposed plan for the London Bridge. This gentleman, however, treats his subject in such a way that only an expert mathematician would attempt to follow him; his book is consequently of little value to the practical **builder, and** the methods usually given for determining **experimentally** the curve of pressure, are of very difficult application, and the results of doubtful accuracy. The formulas for determining the thickness of abutments, as given by Hutton and others, are based upon principles which are not quite practically **correct, and give results too small to be relied upon.**

Influenced by these considerations, the writer has been induced to propose a method for determining the curve of equilibrium, or in other words the line of direction of the pressures, which **he believes to be new, simple, and easy of application.**

If this volume shall **be found to possess no other** merit, **the author can at least claim that it is not a compilation. All the prepositions, with perhaps one or two** exceptions, **have been proved by modes of demonstration** which **he believes to be entirely new, and different from** those employed by Tredgold, **Gregory,** Hutton, and **other** writers **on mathematics or mechanics: he was** so situated at the **time when his attention was first directed to these subjects,** that he could **not procure the works of the standard writers** above referred to for reference, **and was therefore led to** originate principles, and modes **of** demonstration, which subsequent **comparison causes him** to think are different **from any heretofore employed, and in** most instances, **simple and direct. The principle already referred** to of determining the strains of beams by the volumes of geometrical solids, and the deflection and extension of the fibres by a comparison of the areas of plane curves, appears not to have been previously employed; at least the writer has never met with any thing of the kind, as far as his acquaintance with the writings of mathematicians extends.

In the hope, that results which have proved of great value to him may not be entirely useless to others, and in the belief that the theory which he has advanced will explain the phenomena observed in the mutual action of the parts of bridges, and furnish the means of proportioning them upon correct principles; the author submits the result of his labors to the consideration of those who are interested in the theory or practice of bridge construction.

CONTENTS

PART I.

	PAGE
RESISTANCE OF MATERIALS	19
Flexure	22
Beams supported at both ends	24
Inclined beams	26
Strength of short rectangular posts	26
Flexure of columns and posts	32
Resistance of posts to flexure	34
Tension	38
Torsion	39
Forms of equal strength	41
Influence of the vertical forces	43
Relative deflections	48
Strength of particular sections	50
Means of determining the constants	59
WOODEN BRIDGES	63
Horizontal strain	65
Vertical force at any point	65
To find the curve which represents the horizontal strain	71
To find the pressure upon the supports when a beam is framed as a cap upon the tops of several vertical posts, and a weight applied directly over one of the posts	73
Strength of a long beam laid over several supports	76
Effects of counter-bracing	82

CONTENTS.

Inclination of braces.. 84
To determine the strain upon counter-braces.................. 86
To determine the strain upon braces and ties................. 87
Of the strain upon the ties and braces at the centre........... 89
Effects upon the braces and ties which result from the introduction of arch-braces..................................... 91
To determine the strains upon the chords..................... 94
Means of increasing the strength of bridge trusses............ 101
On the maximum span of a wooden bridge..................... 104
Effects of counter-bracing upon an arch....................... 106
Roadway.. 108

CAST-IRON BRIDGES... 110

APPLICATION OF RESULTS.. 114
To determine the strain upon the chords..................... 115
Of the strain upon the lower chord, at the centre............ 116
Strain at the ends of the chords.............................. 116
Of the strain upon the ties and braces....................... 117
Counter-braces.. 120
Lateral horizontal braces..................................... 120
Diagonal braces... 121
Floor beams... 121
Amount of counter-bracing which an arch requires........... 123

EQUILIBRIUM OF ARCHES... 125
To find the thickness of abutments of arches of any kind...... 125
To find the relative length of the joints at different points of an arch, and the line of direction of the pressure............. 136

ILLUSTRATIONS OF PARTICULAR MODES OF CONSTRUCTION............... 141
Foot bridge across the river Clyde............................ 141
Bridge over the torrent of Cismore............................ 142
Bridge across the Portsmouth river........................... 144
Timber-bridge over the river Don, at Dyce in Aberdeenshire... 145
Schaffhausen bridge... 145
Long's bridge.. 146

LATTICE BRIDGES.. 148

IMPROVED LATTICE... 152
Columbia bridge... 155

PART II.

PREFACE... 161
PENNSYLVANIA RAILROAD VIADUCT.................................... 167
Superstructure... 169

CONTENTS.

Bills of materials for one span............................ 171
Bill of castings... 172
Bill of bolts.. 172
Arch suspension bolts...................................... 172
Weight of nuts for one span................................ 173
Estimate of cost   do.................................... 173
Workmanship.. 173
Principles of calculation.................................. 174
Calculation of the strength of the bridge on the supposition that
    the arch sustains the whole weight..................... 175
Strain upon the arch suspension rods....................... 176
Strain upon the counter-bracing produced by the action of the
    arch... 176
Strength of the truss itself without the arch.............. 178
Strain upon the ties....................................... 179
Strain upon the floor beams................................ 182
Strain upon the counter-braces............................. 183
Lateral braces... 184
Strain upon the diagonal braces............................ 185
Resistance to sliding upon the supports.................... 187
Power of resistance, on the supposition that the arches and truss
    form but a single system............................... 188
Estimate of the longitudinal strains....................... 189
General summary.. 192
Strains upon the parts when both systems are united........ 193

COVE-RUN VIADUCT.. 194
   Description... 194
   Bill of materials for one span............................ 195
   Bill of malleable iron.................................... 195
   Wood.. 196
   Recapitulation.. 196
   Calculation... 196
   Estimate of cost.. 197

IRON BRIDGE ACROSS HARFORD-RUN, BALTIMORE..................... 197
   Description... 197
   Bill of materials... 199
   Wood for the whole bridge................................. 200
   Recapitulation of bill of materials....................... 200
   Estimate of cost.. 200
   Estimate of cost of a single track bridge, with two trusses.. 201
   Calculation... 201
   Principle of calculation.................................. 202
   Recapitulation.. 203

LITTLE JUNIATA BRIDGE... 203
   Description... 203

CONTENTS.

Bill of materials for one span.................................. 206
Malleable iron for one span.................................. 207
Nuts.. 207
Summary.. 208
Estimate of cost of one span.................................. 208
Workmanship.. 208
Data for calculation...................................... 209
Calculation of strains...................................... 209
Strain upon the chords.................................... 210
Strain upon the posts.................................... 211
Strain upon the ties...................................... 211
Lateral and diagonal braces................................ 212
Floor beams... 212
Counter-braces.. 213
Strain upon the counter-braces............................ 214
Vertical pressure upon the arch and posts.................. 217
General summary of results................................ 217

SHERMAN'S CREEK BRIDGE, PENN. CENTRAL RAILROAD............ 219
Bill of timber for one span.............................. 220
Bill of counter-brace rods, for one span.................. 221
Arch suspension rods, for one span........................ 221
Floor beam rods, for one span............................ 222
Small bolts, for one span................................ 222
Dimension and data for calculation........................ 222
Calculation of truss without the arches................... 223
Ties and braces... 223
Floor beams... 225
Lateral braces.. 225
Strain upon the knee-braces.............................. 226
Pressure upon the arch.................................. 228
Vertical pressure....................................... 230
Estimate of cost of one span............................. 233
Summary.. 233

RIDER'S PATENT IRON BRIDGE................................ 234
Description... 234
Bill of materials for a single span, 60 ft................ 234
Cast-iron... 234
Malleable iron.. 234
Wood.. 235
Approximate estimate of cost............................. 235
Calculation... 235

CUMBERLAND VALLEY RAILROAD BRIDGE......................... 237
Bill of timber for one span, 186 ft...................... 238
Iron rods... 239
Estimate of cost of one span............................. 239

CONTENTS. 15

Data for calculation..	239
Strain upon the ties	240
TRENTON BRIDGE	242
IRON ARCHED BRIDGE, of 133 feet span.	243
Description	243
IRON BRIDGE OVER RACOON CREEK	247
Bill of materials	248
Estimate	251
Workmanship	251
Data for calculation	252
BALTIMORE AND OHIO RRILROAD BRIDGE	253
Description of details	253
CANAL BRIDGE, PENN. RAILROAD	254
BOILER PLATE TUBULAR BRIDGE	255
ARCHED TRUSS BRIDGE, READING RAILROAD	257
BRIDGE ACROSS THE SUSQUEHANNA	258
Bill of timber	259
IMPROVED LATTICE BRIDGES	260
TRUSSED GIRDER BRIDGES	265

PART I.

RESISTANCE OF TIMBER

AND

OTHER MATERIALS.

CALCULATIONS for the purpose of determining the relations which the dimensions of timbers should bear to the weights which they are required to sustain, are based upon several hypotheses which experience has proved to be correct within the usual practical limits.

The most important of these are—

1. The fibres are susceptible of compression and extension.

2. The degree of extension or compression will be directly as the force by which it is produced.

3. So long as the elasticity remains unimpaired, or so long as the fibres may be considered as perfectly elastic, the force required to produce a given extension will be equal to that which produces an equal compression, and the resistances to these forces will be likewise equal.

These hypotheses will be applied to the most simple case of flexure, which is, that of determining the relations between an applied weight and the dimensions of a timber which are necessary to sustain it when one end is fixed and the other unsupported.

BRIDGE CONSTRUCTION.

Fig. 1.

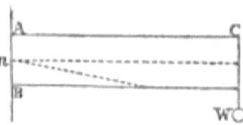

Let AC represent a beam fixed at A and loaded at C with a weight (w), the weight of the beam itself being for the present disregarded—

The substance of the beam is supposed to be entirely uniform throughout, and composed of an assemblage of parallel fibres, all being equally strong.

The effect of the weight w is to stretch the fibres at A and compress those at B. From these points to the interior of the beam the forces gradually diminish, and there must exist some point of the line AB, at which no horizontal force is exerted, and at which the fibres suffer neither extension or compression.

To that line of the longitudinal section which passes through this point, parallel to the direction of the beam AC, has been given the name of the neutral axis, a term which will hereafter be very frequently employed.

The position of the neutral axis will vary with the form of the beam, with the degree of uniformity which it possesses, and with the amount of flexure caused by the load; but in a beam that is straight-grained, rectangular, without knots or flaws of any kind, and not subjected to the action of a weight sufficient to impair its elasticity, it is practically correct to assume the position of the neutral axis in the middle of the section.

Admitting, then, that within the usual practical limits, it is sufficiently correct to assume the position of the neutral axis in the middle of the beam, it is evident that from this line in the direction nA and nB the pressures on the fibres will increase directly as the distance, and if the pressure upon any fibre at the distance $\frac{d}{2}$ be designated by R, the pressure upon any other fibre may be determined from a simple proportion. The total pressure upon the line nB can then be directly determined; for since the pressure upon any individual fibre is as the distance from the neutral axis, it would be represented by the perpendicular erected upon the base ($\frac{1}{2} d$) of a right angled triangle whose altitude is R, and the whole pressure would be represented by the area of this triangle or by $\frac{1}{2} d \times \dfrac{R}{2} = \dfrac{d R}{4}$.

FIG. 2.

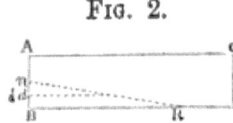

The several forces which act upon the beam may be considered as tending either to cause, or to prevent motion around the point n, and their effects must be ascertained by comparing the products of their intensities by the distances from the point of rotation at which they act.

If, for example, a weight should be applied at the extremity of a lever, its effect would not be represented by the weight alone, but by the weight multiplied by the distance from the fulcrum at which it acts; this product is the moment of the force, and it is these moments, in reference to the axis or point of rotation, and not simply the absolute intensities of the forces, that must be compared in determining the conditions of equilibrium in any system.

Now the weight of any body may be supposed concentrated at its centre of gravity; and, in general, any number of parallel forces may be replaced by a single force called the resultant. In the present case, the pressure of the triangle, which represents the sum of all the forces upon the fibres of the lower half of the section $A\,B$, will be the same, as if a single force equal to its area was applied in the direction of a line passing through its centre of gravity.

As the centre of gravity, or centre of parallel forces of a triangle, is in a line drawn from the vertex to the middle of the base, and at a distance from the latter equal to one-third the length of the bisecting line, it follows that the leverage of the triangle of pressure will be two-thirds of $n\,B$ or $\frac{1}{3} d$; this multiplied by the area of the triangle, (i. e.), by the resisting force along $n\,B$ which we have found to be equal to $\dfrac{d R}{4}$, will give for the moment of this force, in reference to the point n, $\dfrac{d\,R}{4} \times \dfrac{d}{3} = \dfrac{R\,d^2}{12}$.

But the part nA opposes a resistance to extension which is equal to that which the part nB opposes to compression, and, as the moments of these forces are equal, the whole moment of the resisting forces will be expressed by $\dfrac{R\,d^2}{6}$.

The weight, which is represented by w, acts with a leverage equal to the length of the beam, and its moment will therefore be $(w\,l)$:

The equation of equilibrium will therefore be $Wl = \dfrac{R\,d^2}{6}$.

In this equation the breadth of the beam has been regarded as unity, but if it be represented by (b) the equation will become
$$W l = \frac{R\,b\,d^2}{6} \dots\dots\dots 1.$$

The **value of** R must be determined by experiment and will depend upon the kind of material. In general, it has been taken too high, and, as a consequence, the dimensions of timbers deduced **from the** formula which contained it have been too small.

Timber should never be subjected to a strain sufficient to destroy its elasticity, and experiments to determine the value of R should be continued for a considerable length of time.

A weight which even after several months would produce any permanent flexure should be regarded as too great.

When a beam is used as part of a frame, such a value must be given **to** the constant in the proper formula, that only a very slight degree of flexure will take place: the limit, assigned by Tredgold, being one-fortieth of an inch for every foot in length, or $\frac{1}{480}\,l$.

Flexure.

To determine the conditions of flexure, let it be supposed that **a** beam is fixed at one end and loaded at the other, the material being perfectly elastic. **It is** evident, in the first place, that the deflection will be directly as the weight, (i. e.), if a given weight produces a given deflection, twice that weight will produce twice that deflection. Again, the deflection will be proportional to the length and the strain upon the fibres, the last of which is represented by R, and as R contains l in the numerator it will consequently, for these reasons alone considered, be as the square of the length; but, again, the deflection, or which is the same thing, the depression at the extremity, will be directly proportional to the amount by which the fibres are extended or compressed, which is also as the length, and, therefore, the deflection must be as the length cubed.

To make this subject more clear, let w represent the weight suspended at the extremity of a beam whose length is l, and let us ascertain the deflection when l becomes $2\,l$.

FIG. 3.

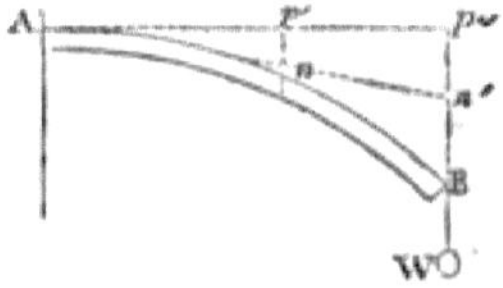

Let $A\,p'\,p''$ represent a horizontal line, and suppose that the action of the weight causes a deflection equal to $p'\,n$, it is obvious that by increasing the length to n', other considerations omitted, the deflection would be $p''\,n' = 2\,p'\,n$.

But if the point of application of w be transferred from n to B, the leverage will be doubled, R will be doubled, and the deflection again doubled from this cause.

Again, the deflection will be as the length of fibre extended.

For example, if a given weight should extend a rod $\frac{1}{10}$ of an inch, the same weight would extend a rod of twice the length $\frac{2}{10}$ or $\frac{1}{5}$ of an inch, and as the deflection must be proportional to this extension, it follows that the deflection must be again doubled from this third cause; hence, the deflection will be directly as the cube of the length.

Lastly, as the deflection is as the value of R, which is $\frac{6\,w\,l}{b\,d^2}$, it will be inversely as the breadth and the square of the depth:

but the quantity of angular motion around the point of support to which the deflection is proportioned is also inversely as the depth, as may be seen by reference to the figure; in which, if $A\,n$ becomes $A\,n' = 2\,A\,n$, the deflection will become $m\,m' = \tfrac{1}{2}\,B\,m$, since, if the leverage of the resistance be doubled, the effect will be reduced one-half. Hence, it follows that the deflection will be inversely as the cube of the depth.

FIG. 4.

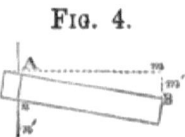

Combining all these results, it follows that the whole deflection will be directly as the weight and the cube of the length, and inversely as the breadth and the cube of the depth—and will be expressed by $\dfrac{W\,l^3}{b\,d^3}$.

If different timbers be required to fulfil the condition, that the deflection shall be equal whatever be the length, we have only to make this expression constant, and determine its value by direct experiment upon the particular kind of timber to be used.

The condition of equal stiffness, however, does not require that the deflection should be equal for every length, but allows it to be in proportion to the length: for example, a beam of 20 feet may be allowed to bend twice as much as one of 10 feet, and the expression modified to suit this case will be $\dfrac{w\,l^2}{b\,d^3}$, a constant quantity for beams of equal stiffness.

By introducing a suitable number for the constant, the equation which expresses its value will determine any one of the four quantities, w, l, b, or d, when the other three are known.

Beams supported at both ends.

When a beam rests on two points of support, and is loaded with a weight applied in the middle, the general circumstances of the case are involved in that which we have considered.

FIG. 5.

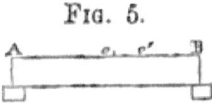

If w represent the weight applied at the centre C, this weight will be transmitted equally to the two points of support at A and B, and the beam may therefore be considered as subjected to the action of two forces, each ($\frac{1}{2} w$) acting with the leverage ($\frac{1}{2} l$) against a fulcrum C.

As the expression for the moment of the resistance is, as formerly, $\frac{R\, d^2}{6}$, and that of the weight $\frac{1}{2} w \times \frac{1}{2} l$, the equation of equilibrium becomes $\frac{R\, d^2}{6} = \frac{w\, l}{4}$, whence $R = \frac{3\, w\, l}{2\, b\, d^2}$, the expression which gives the relative dimensions when a beam is supported at both ends.

If flexure be regarded, the same equation that was obtained in the case of beams fixed at one extremity will be applicable; the only change required is in the value of the constant, for which, see table at the end of this treatise.

When the weight is not in the centre.

In this case, let c represent the distance of the point of application C' from the centre, then the proportion of the weight sustained by B will be determined from the proportion.

$$l : (\tfrac{1}{2} l + c) :: w : \frac{w}{l} (\tfrac{1}{2} l + c).$$

The portion sustained by A will be

$$l : (\tfrac{1}{2} l - c) :: w : \frac{w}{l} (\tfrac{1}{2} l - c).$$

The moments of these forces, in reference to the point C', will be

$$\frac{w}{l} (\tfrac{1}{2} l + c) \times (\tfrac{1}{2} l - c) = \frac{w}{l}\left(\frac{l^2}{4} - c^2\right) = \text{moment of the force at } B.$$

$$\frac{w}{l} (\tfrac{1}{2} l - c) \times (\tfrac{1}{2} l + c) = \frac{w}{l}\left(\frac{l^2}{4} - c^2\right) = \text{moment of the force at } B$$

These moments are equal to each other, and as the resistance of the beam is denoted by $\frac{R\, b\, d^2}{6}$, the equation of equilibrium will be $\dfrac{R\,b\,d^2}{6} = \dfrac{w\,(l^2 - 4\,c^2)}{4\,l}$,

whence, $R = \dfrac{6\,w\,(l^2 - 4\,c^2)}{4\,b\,d^2\,l}$ when the load is not in the middle.

Inclined Beams

The formulæ for horizontal timbers are equally applicable to those which are inclined by taking the horizontal distance between the extreme points, or the cosine of the inclination for the value of l.

FIG. 6.

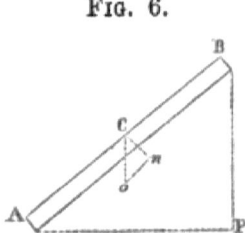

Let the weight w, suspended at C, be resolved into the two components Cn and no; the latter, being parallel to the fibres, will be destroyed by the resistance of the fixed point A. The other, perpendicular to the fibres, is the only one to be considered.

The value of this force is expressed by $Co \times \cos.\,o\,Cn = w \cos. BAP$, and, substituting this value, in the formula, $R = \dfrac{3\,w\,l}{2\,b\,d^2}$, it becomes $R = \dfrac{3\,l\cos.\,BAP}{2\,b\,d^2}$, which differs from the former only in containing $l \times \cos. BAP$, or AP in place of l.

Strength of Short Rectangular Posts.

The investigation of this subject appears to present considerable difficulty, arising from the fact, that the point of application and the direction of the pressure are often indeterminate, and if an attempt is made to express by a formula the conditions of equilibrium, it is necessary to assume data which may only approximate to the probable condition of things in practice. Fortunately, the difficulty is more theoretical than practical; if a post be long and composed of elastic materials, **the condition**, that it shall be of sufficient dimensions to resist flexure with a given applied weight, brings the question into a tangible form, and the solution is simple; on the other hand, if the post is short, the **usual limit** of the weight, which is generally one **tenth of that** which would be required to crush the material, is amply sufficient to compensate for any variation in the point of application, or line of direction of the pressure that may be produced by unequal settling or other causes.

If we attempt to continue the subject, as heretofore, by establishing relations between the moments of the acting and resisting forces, the first step must be the determination of the position of the neutral axis; as no comparison of moments can be made until we know the point or axis of rotation to which they are referred.

In posts placed vertically and loaded at their upper extremities the position of the neutral axis can no longer be assumed in the centre, but will vary greatly according to the amount and point of application of the weight, and the degree of flexure that is supposed to have been produced.

It appears reasonable to conclude, that when a post is uniform in composition and acted upon by a force applied exactly in the direction of the axis, all parts of the central cross section are subjected to nearly equal degrees of pressure; but if the weight be applied at either side of the axis, the compression no longer continues uniform but becomes greatest on that side towards which the pressure is applied.

Fig. 7.

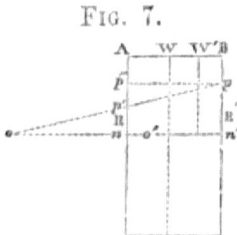

Let $A\ B\ C\ D$ represent the longitudinal section of a rectangular post of some stiff material: when the weight is applied at w every part of the cross section ($n\ n'$) may be considered as subjected to equal pressure, and if R and R' represent the forces acting upon the fibres at n and n', these forces may be represented by some portion of their lines of direction, as $n\,p''$ and $n'\,p$.

In the present case, these forces are supposed equal, and the line $p\,p''$ will be parallel to $n\,n'$; the point of intersection, which determines the distance of the neutral axis, will therefore be infinite.

If w be applied at W' the pressure at n' would become greater than at n, and if $n\,p'$ and $n'\,p$ represent the relative magnitudes of the forces at n and n', the line $p\,p'$ will be inclined to $n\,n'$ and must intersect at some point O, and, consequently, the distance of the neutral axis will become finite and could be determined if we knew the relative values of the forces at n and n'.

As w approaches B, the differences of the forces at n and n', which, for brevity, will be called R and R', will become greater, and O will approach n.

After the post yields laterally, the fibres along $A\,C$ will be extended, and, before it arrives at this point, there must be a certain magnitude of the weight, or a certain position of its line of application, that will cause neither extension or compression along $A\,C$, which will accordingly become the neutral axis.

With a still greater weight, the fibres along $A\,C$ being extended, and those along $B\,D$ compressed, the neutral axis must be within the post at some point O'.

If the post is long, and the pressure be supposed still to increase, the elasticity of the timber being unimpaired, O' will approach very near the centre. Lastly, a still greater increase of weight, and consequent flexure, will destroy the elasticity, and the position of the neutral axis will then depend on the relative powers of the fibres to resist the crushing or extending forces.

Let it be assumed, that the direction of the weight coincides with the edge of the post, and that $A\,C$ suffers neither extension or compression, the neutral axis will be at n, and R, as formerly, representing the maximum force, which will be at n', the resistance will be expressed by $(n\,n' = b) \times \dfrac{R}{2}$, its moment will be $\dfrac{b\,R}{2} \cdot \tfrac{2}{3} b = \dfrac{b^2\,R}{3}$. The moment of the weight being

($b\,w$), the equation of equilibrium becomes $\dfrac{b^2 R}{3} = b\,w$, or $W = \dfrac{bR}{3}$.

FIG. 8.

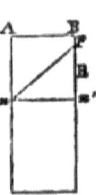

When the weight is applied directly in the axis, every part of the section $n\,n'$ sustains an equal portion.* We have therefore $w = b\,R$. Hence it appears, that a post will sustain three times as much, when the weight is applied along the axis, as it will when the line of direction coincides with one of the sides, provided, the dimensions are such, that flexure can take place only in the direction of b.

Tredgold, in his treatise on cast iron, page 234, makes the resistance one-fourth when the weight acts on one edge of a block. This requires, that the neutral axis should be within the rectangle at a distance from the line of pressure equal to $\frac{3}{4}$ the breadth, and that, beyond the axis, the parts oppose no resistance, a supposition, which, we think, is less nearly correct than that which we have assumed.

---

* This would be true in all cases were the material perfectly unelastic; but if the lateral cohesion of the fibres be not so great as to prevent any motion, some variation in the degree of pressure upon different parts of the cross section must ensue. This will certainly be the case when the support is very short in proportion to its width: for example, a weight applied at A would produce a tendency to flexure in the direction of the dotted lines, and then the pressure at p would be greater than at n or n'. This objection is of no practical importance, as supports are always too long to allow of flexure in this way.

FIG. 9.

We will continue the hypothesis, that the neutral **axis is on** one side and the direction of **the** pressure on the other.

When the line of direction of the weight coincides with the axis of a column, the strength will be ⅙ as great as when it coincides with one side.

FIG. 10.

Let AB represent the line of direction of the weight, CD = the neutral axis, R = strain upon n', the strain upon any point p would be represented by a perpendicular through that point terminated by the oblique plane An, the whole pressure would consequently be represented by the semi-cylinder Ann'. The vertical line, through the centre of gravity, passes at a distance from $n = \frac{5}{4}$ radius.*

---

* As the centre of gravity of this solid is not given in any mathematical work to which the author has access, he thinks it proper to explain the method by which he has obtained the distance $\frac{5}{4}r$.

*To find the volume and **centre** of gravity of a semi-cylinder cut off by an oblique plane passing through the edge of the base.*

FIG. 11.

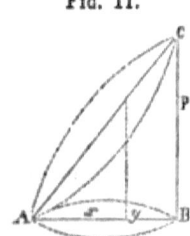

Let r = radius, x = any abscissa, y the corresponding ordinate of the circle.

Then $2r : x :: R : \dfrac{R}{2r} x$ = perpendicular of elementary rectangle

RESISTANCE OF MATERIALS. 31

The moment will therefore be $\dfrac{\pi r^2}{2} \cdot R \cdot \tfrac{5}{4} r = \tfrac{5}{8} \pi r^3 R$.

The moment of the weight, acting with a leverage $2 r$, is $2 w r$. The equation of equilibrium is $\tfrac{5}{8} \pi r^3 R = 2 w r$, or $\tfrac{5}{16} \pi$

$\dfrac{R}{2 r} x \cdot 2 y = \text{area} = \dfrac{R}{r} x y$. $\dfrac{R}{r} x y d x = $ elementary solid.

$\dfrac{R}{r} y x^2 d x = $ moment of elementary rectangle.

But from the equation of the circle we have $y = \sqrt{2 r x - x^2}$ $\int \dfrac{R}{r} x y d x = \dfrac{R}{r} f (2 r x - x^2)^{\tfrac{1}{2}} x d x$ make $(r - x) = z$, $d x = - d z$, $2 r x - x^2 = r^2 - z^2$. Substitute these values we obtain $\dfrac{R}{r} f (2 r x - x^2)^{\tfrac{1}{2}} x d x = \dfrac{R}{r} f (r^2 - z^2)^{\tfrac{1}{2}} (r - z) (- d z) = - \dfrac{R}{r} [f (r^2 - z^2)^{\tfrac{1}{2}} r d z + f (r^2 - z^2)^{\tfrac{1}{2}} z d z]$. The first of these integrals taken between the limits $+ r$ and $- r$ is the area of a semi-circle and is consequently equal to $\dfrac{\pi r^2}{2}$ hence the value of the first term becomes $\dfrac{\pi r^3}{2}$.

The second term becomes $\dfrac{(r^2 - z^2)^{\tfrac{3}{2}}}{3}$ (see demonstration of ungula, Prob. 14), and is equal to 0 when $z = + r$, or $z = - r$; it therefore disappears, and the **volume of the** solid becomes $- \dfrac{R}{r} \cdot \dfrac{\pi r^3}{2} = - \tfrac{1}{2} \pi r^2 R$; a result which is evidently correct, since the volume is half that of the cylinder $\pi r^2 R$. To find the distance to the centre of gravity we must divide the integral of $\dfrac{R}{r} y x^2 d x$ by the volume. Making similar substitutions to those used in finding the volume, we obtain $\dfrac{R}{r} f y x^2 d x = \dfrac{R}{r} f (r^2 - z^2)^{\tfrac{1}{2}} (r - z)^2 (- d z) = \dfrac{R}{r} f (r^2 - z^2)^{\tfrac{1}{2}} (r^2 - 2 r z + z^2) (- d z) = \dfrac{R}{r} [r^2 f (r^2 - z^2)^{\tfrac{1}{2}} d z - 2 r f (r^2 - z^2)^{\tfrac{1}{2}} z d z + f (r^2 - z^2)^{\tfrac{1}{2}} z^2 d z]$.

The first of these integrals is a semi-circle, hence $r^2 f (r^2 - z^2)^{\tfrac{1}{2}} d z = r^2 \dfrac{2 \pi r^2}{2} = \pi r^4$.

The second, $f (r^2 - z^2)^{\tfrac{1}{2}} z d z$, as we have seen, becomes $= 0$.

The third, $f (r^2 - z^2)^{\tfrac{1}{2}} z^2 d z$, is proved, in the problem of the ungula, to be $\tfrac{1}{4} r^2 f (r^2 - z^2)^{\tfrac{1}{2}} d z$, which, between the limits $+ r$ and $- r$, becomes $\dfrac{\pi r^4}{8}$. (See note to Prob. 14.)

$r^2 R = w$. But when the pressure coincides with the axis, we have $\pi r^2 R = w$; hence, the strength in the two cases will be as 5 to 16.

Flexure of Columns and Posts.

It is evident that if a column be perfectly cylindrical, and the direction of the weight coincide exactly with the axis, flexure cannot take place; but if the weight be sufficient the fibres will yield by crushing. Flexure therefore must result from some obliquity in the line of direction of the force.

Fig. 12.

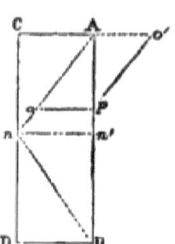

Let us take the most unfavorable case that would probably occur in practice, as it is that which gives the greatest diameter and is consequently the most safe.

Let A be the point of application of the weight, AB its line of direction, CD the position of the neutral axis at the instant

The whole expression therefore becomes $\dfrac{R}{r}\int y x^2 dx = -\dfrac{R}{r}\left(\dfrac{\pi r^4}{2} + \dfrac{\pi r^4}{8}\right) = -\tfrac{5}{8}\pi R r^3$ and $\dfrac{\dfrac{R}{r}\int y x^2 dx}{\dfrac{R}{r}\int y x\, dx} = \dfrac{-\tfrac{5}{8}\pi R r^3}{-\tfrac{1}{2}\pi R r} = \dfrac{5}{4}r$.

Hence, the line through the centre of gravity, perpendicular to the base, passes at a distance of $\tfrac{5}{4} r$ from the centre; the centre of gravity will be found in this line, and also in the line drawn from A to the middle point of BC; hence, it will be at their intersection, and its height above the base can be found by the proportion $2r : \dfrac{R}{2} :: \tfrac{5}{4} r : \tfrac{5}{16} R$.

RESISTANCE OF MATERIALS. 33

of flexure. The force at A, in the direction AB, acts with a leverage nn', and tends to produce rotation around the point n.

Join An, and let the weight (w) be represented by the portion AP of its line of direction. By constructing the parallelogram of forces on An and An', it will be evident that, in consequence of the obliquity of the line An, there will result a horizontal component (Ao'), the magnitude of which will b proportional to the cosine of the inclination Ann'. If A be fixed, as it always is in columns, so that it cannot move in the direction Ao', the force Ao will be transmitted to n, and its horizontal component, $=Ao'$, will produce a cross strain upon the middle of the column at n.

The reaction of the point B, upon which alone the column is supposed to rest, produces a force $= w$, and acting in an opposite direction, its horizontal component at n will be equal to Ao', and the fibres at n will be subjected to a cross strain resulting from the actions of these two forces equal to $2Ao'$.

The vertical component is resisted by the strength of the material, and, as it is the horizontal force alone which tends to produce flexure, this alone will be considered.

As the column is fixed at the points A and B, and subjected to a cross strain at n, it is in the condition of a beam supported at both ends and loaded in the middle, and, therefore, the conclusions at which we arrive in the former case are, with slight modification, applicable here.

It was shown, that beams to be equally stiff must be proportional to $\dfrac{wl^2}{bd^3}$, but in the case of a cylinder $b=d$, and the expression becomes $\dfrac{wl^2}{d^4}$, in which d represents the diameter.

The expression for the strain $\left(\dfrac{wl^2}{d^4}\right)$ shows that it is directly as the weight and square of the length, and, inversely, as the fourth power of the diameter; * but the weight does not, in

---

* That the strain is inversely as the fourth power of the diameter may also be shown by the following considerations:

this case, represent the pressure on the top of the column, but the cross strain at the middle.

The value of the constant, in rectangular beams, was determined by the condition, that the flexure should be $\frac{1}{480}$ of the length, or $\frac{1}{40}$ of an inch per foot. The same constant would give for a column 10 feet high a deflection of $\frac{1}{4}$ of an inch. If this be considered too great, the constant must be increased: but it must be remembered that this is the maximum deflection, on the supposition, that the weight is thrown altogether upon one side of the column, the most unfavorable case that can occur; it is therefore probable that no change in the value of the constant is required.

When the height of a column does not exceed about nine times the diameter, it is found, that the fibres will crush before they will yield laterally, and the strength will therefore be proportional to the area of the section, or d^2; we have in this case, $d^2 R = w$.

Resistance of Posts to Flexure.

The ordinary formula for the stiffness of beams, supported

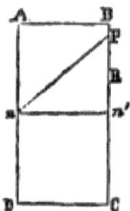

If AD be supposed to represent the neutral axis, and R the maximum strain upon the fibres BC; the pressure upon any part of the section nn' would be represented by a perpendicular to nn' terminated by the oblique plane pn. The solid $nn'p$, whose altitude R is constant, and whose base is equal to the area of the section, will therefore represent the pressure upon nn', and will be proportional to d^2. The leverage, being the distance from n to the perpendicular, through the centre of gravity, will also be proportional to d, and therefore the strength of the cross section would be in proportion to d^3, and the strain inversely as d^3. The strain will also be as the deflection, which, as in the case of horizontal beams, can be shown to be inversely as the diameter; hence, combining all these results, the strain will be inversely as d^4.

at the ends and loaded in the middle, is
$$w = \frac{b\,d^3}{c\,l^2},$$
in which (w) represents the weight which produces a given deflection, $b =$ breadth in inches, $d =$ depth in inches, and $l =$ length in feet; c is a constant, to be determined by substituting the values of the other quantities in the equation.

In making experiments to determine the constant from this formula, it is necessary to observe very accurately, **both the weights and the deflections** produced by them, and then, by means of a proportion, find the value of (w), which will produce the deflection required to be substituted in the formula.

In reflecting upon the circumstances connected with the flexure of beams, **the** writer conceived the idea **of** deducing an expression for the weight which a post would support from the ordinary **formula for** the stiffness of a **horizontal beam, by the** following **considerations.** If a beam is **bent by an applied** weight, there will be a tendency, from the elasticity of the material, to recover its form when the weight is removed; but if the ends are fastened by being placed between resisting points, so that the piece cannot recover its shape, there must be a horizontal force caused by the reaction of the material; and this force is such, that if the beam were placed in **a vertical** position and loaded with a weight equal to it, the **deflection should be the** same **as that of the horizontal beam,** and consequently **the extreme limit of the resistance of the post to** flexure **would be** determined.

To ascertain the force which **is exerted** by the reaction of a bent beam in the direction of the **chord** of the arc.

FIG. 13.

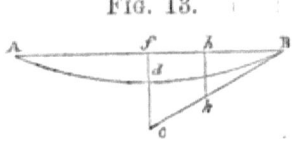

Let $A\,B$ represent a beam, supported **at the** ends and loaded with a weight (w) applied at the middle point.

$d =$ deflection caused by the applied weight.

$B\,C =$ tangent of curve at B.

If the weight be removed, the reaction of the beam will cause it to regain its original figure if not resisted by a pressure at the ends. The force of this reaction will be proportional to the degree to which the **fibres are** strained, and as the strain upon the fibres is nothing at the ends A and B, and increases **uniformly to the middle** point, the force of reaction will be in **the same proportion,** and the point of application of the resultant **of the whole** of the reacting forces will correspond to the **centre of** gravity **of a** triangle, whose base is Bf; it will consequently be at a distance from $B = \tfrac{2}{3} \, Bf$.

The effect of this resultant acting at a distance $\tfrac{2}{3} \, B f$, **must be the same** as the weight $\left(\dfrac{w}{2}\right)$ acting **at** a distance Bf, and **must** consequently be in proportion to $\dfrac{w}{2}$ **as** $3 : 2$. The value of the resultant is therefore $\dfrac{3\,w}{4}$.

The line of direction of the pressure **at** B being **the tangent** BC, the **force** of reaction at h may be considered as applied at the point k of its **line** of direction, and as $k\,h\,B$ and $Cf B$ are **similar** triangles, $Cf : fB :: \tfrac{3}{4} \, w :$ horizontal pressure at $B = \tfrac{3}{4}\,w \times \dfrac{fB}{fC} = \tfrac{3}{4}\,w \dfrac{\tfrac{1}{2}l}{2\,d} = \dfrac{3\,w\,l}{16\,d}$. Representing this force by P we have

$$P = \frac{3\,w\,l}{16\,d}.$$

As the deflection **of** a beam within the elastic limits is always **in** proportion to the weight, if $(w') =$ the weight that will produce a deflection equal to **unity,** the deflection (d) will require a weight $= (d\,w')$, and by substituting this value in the equation, **we** find

$$P = \tfrac{3}{16} \cdot \frac{d\,w'\,l}{d} = \tfrac{3}{16}\,w'\,l.$$

In this expression (d), which represents the deflection, has disappeared, and as (w') is a constant quantity for the same beam, representing the weight that produces a deflection equal to the unit of measure, it follows, that P is the same with every weight and every degree of deflection within the elastic limits.

This result seems, at first view, to be contrary to fact; it would appear that if the weight is increased, the horizontal strain should be increased in the same proportion; but when it is remembered, that the deflection increases with the weight, and that the former diminishes the value of P in precisely the same proportion that the latter increases it, the difficulty vanishes, and the reason why P should be constant for the same beam becomes obvious.

The practical importance of this result is very great, as it furnishes the means of obtaining a formula, which will give at once the extreme limit of the resistance to flexure, or the weight which, applied to a post, will cause it to yield by bending.

As the formulæ used by Tredgold are calculated for a deflection of $\frac{1}{40}$ of an inch to one foot, or $\frac{1}{480}$ of the length, the weight which would cause a deflection of 1 would be w $\left(1 \div \frac{l}{480}\right) = \frac{480\,w}{l}$, and by substituting this value for w' in the equation

$$P = \tfrac{3}{16}\,w'\,l,$$

we find $P = 90\,w = A$.

But from the ordinary formula for the stiffness of a beam supported at the ends, we have

$$w = \frac{b\,d^3}{c\,l^2}. \quad \text{Therefore } P = \frac{90\,b\,d^3}{c\,l^2} = B.$$

The expression $P = 90\,w$ shows that the extreme limit of the strength of any post whatever, of any length, breadth, or depth, or of any kind of material, is ninety times the weight which causes a deflection of $\frac{1}{480}$ of the length.

The second expression, $P = \dfrac{90\,b\,d^3}{c\,l^2}$, will give the value of P directly, without first knowing the weight required to cause a given deflection in a horizontally supported beam. In this expression, b = breadth in inches, d = depth or least dimension in inches, l = length in feet, and c = a constant to be determined by experiment for each species of material.

The value of c for white pine is ·01. By substituting this value, we find $P = \dfrac{9000\,b\,d^3}{l^2}$, a remarkably simple formula, which gives the extreme limit of the resistance to flexure of a white pine post.

The same expression may be employed to determine the constants used in the ordinary formula for the stiffness of beams. For this purpose let the equation $P = \dfrac{90\,b\,d^3}{c\,l^2}$ be transposed, which will give $c = \dfrac{90\,b\,d^3}{P\,l^2}$. Find P by applying a string to a flexible strip of the material to be experimented upon, in the manner of a chord to an arc, and ascertain the tension on the chord with an accurate spring balance. It will be found that, whether the strip be bent much or little, the tension on the chord, as shown by the spring balance, will be constant, and this tension, in pounds substituted for P, will give the value of c without requiring, as is necessary with other formulæ, an observation of the deflection.

Experiments made upon these principles with strips of white pine, **yellow pine, and** white oak, 5 feet long, 1¼ inches wide, **and** ¼ **inch** deep, gave the following results:—

The observed tensions were,

        White Pine,   7¼ lbs. value of $c = \cdot 0097$
        Yellow Pine, 6½ "   "   " $= \cdot 0108$
        White Oak,   6¾ "   "   " $= \cdot 0104$

As the stiffness is inversely as these **constants, it** follows that white pine is stiffer than yellow pine **or** oak. The experiments of **Tredgold** give similar results.

Tension.

When a force is applied in **the** direction of the axis of a suspension rod, the resistance is directly proportional to **the** area of the section; and, consequently, it is only necessary to multiply this area by the number expressing the resistance of a square inch. As metals are the only substance well suited to resist tensile strains, we find that they are almost exclusively employed for **this** purpose, and generally in such lengths, when compared with their **diameters, that the directions of the strains will always coincide very nearly with the axis.**

If, however, in extreme cases, it should happen, **that the** weight can be thrown altogether on **one** side, then it is necessary to increase **the area** of the **section**; the amount of which increase will be determined upon **similar principles to those which apply in the case of columns.**

Torsion.

The resistance which is opposed by a shaft to **a twisting force** is called resistance to torsion.

The following method of investigation is similar to that pursued by Tredgold.

If the shaft be of the form of a rectangular plate, we may suppose it to be supported at the corners A and B, and weights suspended from each of the other corners C and D; **the strain** produced by loading it in this manner would be similar to the twisting strains in shafts. As the weights at **the** four corners are supposed equal, there would be an equal tendency to break **at** the same time along the diagonals $A B$ and $C D$, or along some other lines, at which the resistance might be less; **but,** before fracture, **one of** these lines will **serve as a fulcrum for** forces acting at the **extremity of the other.**

FIG. 14.

To determine the **line** of fracture when the **material is uni**form **m** composition and equally tenacious in every direction, let z represent the length $A B$ of the **line** of least resistance, b = the breadth of the plate $B C$. Then $A C = \sqrt{z^2 - b^2}$, and if y represent the perpendicular Cp, we have $z : \sqrt{z^2 - b^2} :: b : y = \dfrac{b}{z} \sqrt{z^2 - b^2}$.

The weight acting at the point C, its effect will be in proportion to the leverage Cp, and the resistance of the section, the depth being constant, will be as its length z. The line of least resistance is that at which z will be least when compared with y, or when $\dfrac{y}{z}$ is a maximum. We have therefore $\dfrac{\sqrt{z^2-b^2}}{z^2}$, which is a maximum when $z = b\sqrt{2}$, or $AC = BC$.

The case which has been considered is the most simple, and applies to the resistance of a uniform material such as cast iron.

Having found the length of z, it is only necessary to substitute its value for the breadth, in the ordinary formulæ for the resistance to a cross strain, in order to determine the strength.

This calculation is evidently founded on the supposition, that the resistance is the same in every direction, as is the case in cast iron and similar materials; but, in shafts composed of a fibrous substance, the line of least resistance would have a different direction, depending upon the relative proportion between the lateral and longitudinal cohesions. It is not necessary, however, that we should proceed to the examination of this case, since the only questions of practical importance are:

What is the resistance to flexure within the elastic limits, and what degree of angular motion will be produced by a given flexure?

The angle of torsion may be deduced from the following considerations. Let (e) be the extension which the material will bear without injury when its length is unity. This extension must obviously limit the movement of torsion, or the angle of torsion; but, since the line of greatest strain in a bar of greater length than the diagonal of the square of its base is in the direction of the diagonal of a square, the whole extension would be in proportion to the length of a line wrapped around the bar at an angle of 45° with the axis, and would therefore be equal to ($le\sqrt{2}$). The arc described, or the angular motion which this extension will allow, is equal to the diagonal of a square, of which this extension is the sides, or $1 : \sqrt{2} :: le\sqrt{2} : 2le$, as is obvious by reference to the figure; the extension of the diagonal $= n$, admitting a degree of angular motion $= m$.

Fig. 15.

$2\,l\,e$ represents the arc described in feet, or $24\,l\,e =$ the arc in inches. But if $a =$ the number of degrees in an arc, and $\dfrac{d}{2}$ its radius, $\cdot 0174533$ being the length of an arc of one degree when its radius is unity, we have $24\,l\,e = \dfrac{a\,d}{2} \times .0174533$, or $a = \dfrac{2750\,l\,e}{d}$.

That is, the angle of torsion (a) is as the length and extensibility of the body directly, and inversely as its diameter.

The value of e for cast iron is $\frac{1}{1204}$, hence $a = \dfrac{2\cdot 284\,l}{d}$.

The value of e for malleable iron is $\frac{1}{1400}$, hence $a = \dfrac{1\cdot 965\,l}{d}$.

*Forms of equal strength for **beams** to resist cross strains.*

Whatever may be the **form of** the beam, it **is** always **necessary** that the area **of the** section resting upon the points **of** support should **be** sufficient to resist the force of detrusion, or that which tends to crush the fibres in a direction perpendicular to their length. This resistance is directly proportioned to **the area,** and if w represent the weight at the point of support, R the resistance per square inch, $b' =$ breadth, and d' the depth; **then,** the dimensions at **the** end must never be less than will be given **by the** equation $w = R\,b'\,d'$.

The practice of other writers has been to omit the consideration of this force in determining the forms of equal strength.

The following results, obtained by omitting for the present the consideration of the detrusive force, agree with theirs.

PROPOSITION 1. *If a beam be supported at the ends, and the load applied at any point between the supports, the extended side being straight, the form of the compressed side will be that formed by two semi-parabolas.*

FIG. 16.

Consider C, the point of application of the weight, as a fulcrum.

Let w represent the portion of the weight sustained by A.
$x =$ the distance to any section whose depth is y.

The strain will be $w\,x$, the resistance proportional to the square of the depth will be y^2; hence $y^2 = w\,x$, which is the equation of a parabola.

PROP. 2. *If the depth be constant, the horizontal section will be a trapezium. For in this case* $w\,x = d^2 y$, $x = \dfrac{d^2}{w} y$: *or* y *is proportional to* x, *which is the property of a triangle.*

FIG. 17.

PROP. 3. *When a beam is regularly diminished towards the points that are least strained, so that all the sections will be circles, or other similar figures, the outline should be a cubic parabola.*

For, in this case, if y represent the diameter, or side of the section, the resistance will be as y^3, and the equation of moments will be $w\,x = y^3$.

The same figure is proper for a beam fixed at **one end** and the force acting at the other.

RESISTANCE OF MATERIALS.

PROP. 4. *If a weight be uniformly distributed over the length of a beam supported at both ends, and the breadth be the same throughout, the line bounding* **the compressed** *side should be a semi-ellipse when the lower side is straight.*

FIG. 18.

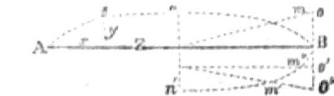

The reaction of the points of support, A and B, may be considered as two forces acting upwards, whilst the uniform weight acts downward; these two classes of forces are therefore opposed to each other, and the strain at any point will be proportional to the difference of their moments.

Call $l = \frac{1}{2} AB.$ $w =$ weight on A.

The moment of w at the distance x will be $w\,x$.

The portion of the weight on x will be $\dfrac{w\,x}{l}$, its moment

$$\frac{w\,x}{l} \cdot \frac{x}{2} = \frac{w\,x^2}{2\,l}.$$

Hence, $w\,x - \dfrac{w\,x}{2\,l} = y^2 \quad \therefore w\,(2\,l\,x - x^2) = 2\,l\,y^2.$

To refer the curve to the centre C, make $l - x = z$, whence $2\,l\,x - x^2 = l^2 - z^2$, and by substitution we have $w\,l^2 - w\,x^2 = 2\,l\,y^2$, or $2\,l\,y^2 + w\,x^2 = w\,l^2$, which is the form of the equation of an ellipse referred to the centre.

Influence of the vertical forces.

The forms which are here given for beams of equal strength correspond with the results obtained by all who have written upon the subject of the strength of materials. But that they are not strictly correct can be readily proved. They have been obtained by directing attention **only** to the horizontal forces which produce longitudinal **strains upon the** fibres of the beam. But another force (the existence and effects of which will be more fully considered when we treat of wooden bridges), appears to have been disregarded. Dr. Young alludes to a force which he calls detrusion, the effect of which is to crush across the fibres close to a fixed point, but no allusion is made, either by him or any other writer, (as far as our information extends,) to the existence of a force acting transversely on the fibres at any other point. It will be shown hereafter, that when the beam is uniformly loaded, the vertical force in the centre is nothing, and it increases uniformly to the ends, where it is equal to half the weight upon the beam. Consequently, if the breadth of a beam is constant (Fig. 18), the true figure of equal strength will not be $A\,s\,n$ but $o\,m\,n$, in which the area of $B\,o$ must be sufficient to resist detrusion at the point B, and $o\,m\,c$ must be a straight line.

If the beam was a solid of revolution, $a\,s\,n$ or $n\,m\,B$ would be a cubic parabola, and $c\,m\,o$ a common parabola.

If, instead of being uniformly distributed, the weight were applied entirely at the centre, the form of equal strength would be determined by the intersection of $n'\,m'\,B$ with the side of the rectangle $C\,o'$, and would be $n'\,m''\,o'$.

Lastly, if there should be both a uniform load and a weight applied at the middle, the figure to resist the vertical forces would be a triangle placed upon a rectangle (a trapezoid), and the form of equal strength in this case would be $n'\,m'\,o''$.

PROP. 5. *If a beam uniformly loaded and depth constant, be supported at the ends, the outline of the breadth should be two parabolas.*

FIG. 19.

The strain upon the section y is $w\,x - \dfrac{w\,x^2}{2\,l}$. The resistance, since the depth is constant, will be proportional to y.

Hence, $\left(w\,x - \dfrac{w\,x^2}{2\,l}\right) = y$.

Substitute $l - z$ for x, to refer to the centre C, it becomes $w\,(l^2 - z^2) = 2\,l\,y$.

Make $z = o$, we have for the ordinate $nc = y = \dfrac{w\,l^2}{2\,l}$.

To refer the curve to the point n, we must make $y' = nc - y$, or $y = nc - y' = \dfrac{w\,l^2}{2\,l} - y'$; hence, $\dfrac{w\,(l^2 - z^2)}{2\,l} = \dfrac{w\,l^2}{2\,l} - y'$.

Reducing, we have $y' = \dfrac{w}{2\,l}\,z^2$, which is the equation of a parabola.

PROP. 6. *If a beam is fixed at one end only, the breadth constant, and the weight uniformly distributed, the form of equal strength will be a triangle.*

FIG. 20.

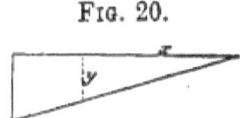

The weight on x is $\dfrac{w\,x}{l}$, the moment is $\dfrac{w\,x^2}{2\,l}$, the moment of resistance y^2.

Hence, $y^2 = \dfrac{w\,x^2}{2\,l}$, or $y = \sqrt{\dfrac{w}{2\,l}} \cdot x$, the equation of a straight line.

If the weight be all at one end we have $y^2 = wx$, a parabola.

If the sections be similar figures $y^3 = wx$, a cubic parabola.

PROP. 7. *The form of a suspension rod of equal strength is determined by* the equation $x = 2\,a\,\log.\,y$.

FIG. 21.

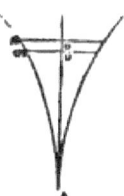

Let w = weight suspended at A
P = weight of any portion of the rod $A\,o$
$x = A\,o,\ y = o\,m,\ dx = o\,o',\ y + dy = o'\,n$
a = resistance of material per square inch.
Then, $(w + P) = a\,y^2$
$y^2\,dx$ = solidity or weight of portion $o\,o'$
$w + P + y^2\,dx = a\,(y + dy)^2$ (subtracting the first equation) $y^2\,dx = 2\,a\,y\,dy$

$$dx = 2\,a\,\frac{dy}{y} \qquad x = 2\,a\,\log.\,y.$$

PROP. 8. *The strength of two similar cylinders, or other solids of the same material, supported at the ends and strained by their own weights, will be inversely as their like dimensions.*

The strength of a single cylinder would be proportional to $\dfrac{d^3}{w\,l}$. Let another similar cylinder be n times the dimension of the first, its weight would be as $1 : n^3$, and its strength $\dfrac{(n\,d^3)}{n\,l \times n^3\,w} = \dfrac{d^3}{n\,w\,l}$; hence, the strength of the first would be to the strength of the second as $n : 1$. If the span of a bridge be doubled, even if the dimensions of all the parts be increased in the same proportion, the strength will only be one-half.

PROP. 9. *The parabolic beam of equal strength contains two-fifths less material than the circumscribing cylinder.*

FIG. 22.

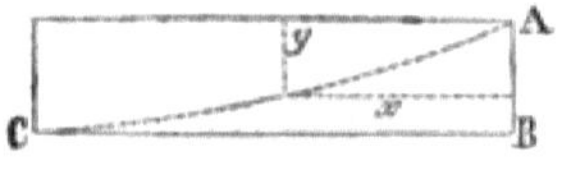

From the equation of the curve we have $y^3 = x$. Hence, the element of the surface $A\,B\,C = x\,dy = y^3\,dy$, and the area $= \int y^3\,dy = \dfrac{y^4}{4} = \tfrac{1}{4}\,x\,y.$

The volume is equal to the surface multiplied by the circumference described by the centre of gravity.

$\dfrac{f x y \, dy}{f x \, dy}$ = distance of centre of gravity = $\frac{4}{5} y$

$2 \pi \cdot \frac{4}{5} y \cdot \frac{1}{4} x y = \frac{2}{5} \pi y \quad x = \frac{2}{5}$ (cylinder generated by revolution of $A\,C$).

Prop. 10. *When a plate is supported at two edges, and a weight applied at the centre, the weight of the plate* **itself not** *being considered, the strength is constant whatever be* **the area.**

Let the plate be rectangular.

Fig. 23.

Take the line $p\,p$, passing through the centre, as the line of fracture. Call the sides a and b. Let $n\,b$ = distance to centre of gravity of each half, regarding the point of application of the weight as a fulcrum.

Then, $\dfrac{w}{2} \cdot n\,b = a\,R$, or $R = \dfrac{w\,n}{2} \cdot \dfrac{b}{a}$, which is a constant as long as the ratio of b to a is constant.

The same is true, if the fracture be supposed to take place along any oblique line, for if the plate be increased or diminished, the **lines** $w\,o$ **and** $A\,B$, which express the leverage of the weight and the resistance, will always bear the same ratio. (Fig. 25.)

Fig. 24.

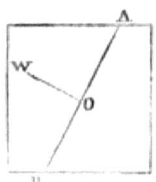

When the weight is uniformly distributed,

$w = \frac{1}{2} a\,b, \quad w \times n\,b = \frac{1}{2} n\,a\,b^2 = a\,R, \quad R = \dfrac{n}{2} b^2,$ or the strain is proportional to the area of the plate. Hence, where there is no applied weight, the strength of the plate diminishes in proportion as the area increases.

RELATIVE DEFLECTIONS.

PROP. 11. *To find the deflection of a rectangular beam supported in the middle, and uniformly loaded over its length.*

FIG. 25.

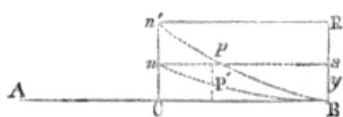

Let AB be the beam, C the fulcrum, $x =$ distance of any perpendicular (y) from the extremity B.

When the weight is at the extremity, the strain upon any section will be as the distance x, and will be represented by wx, but the deflection will be not only as the strain, but as the distance from B; hence, it will be proportional to wx^2, or, if $y = wx^2$, it is evident that y corresponds to the abscissa of a common parabola whose ordinate is x, and the whole deflection equal to the sum of these abscissas will be represented by the area $BCn' = \frac{1}{3}$ rectangle $BCn'R = (A)$.

When the weight is uniformly distributed, the strain upon any section will be in proportion to the weight and distance from B.

Let x be any distance, then $l : w :: x : \dfrac{wx}{l} =$ weight on the part x, $\dfrac{wx}{2l} \cdot x =$ moment to which the strain or extension of the fibres will be proportioned. The deflection, being as the strain and distance from B, will be $\dfrac{w}{2l} x^3$.

If, then, $y' = \dfrac{w}{2l} x^3$, we perceive that y' is the abscissa of a cubic parabola of which x is the ordinate.

The area $B\,Cn = \frac{1}{4}$ of rectangle $B\,Cn\,s = \frac{1}{8} B\,Cn'R =$ the deflection.

Hence, **the deflection in the two cases will be** as $\frac{1}{3}$ to $\frac{1}{8}$, or as 8 to 3.*

Prop. 12. *The deflection of a beam supported* **at the ends** *and uniformly loaded will be to the deflection of the* **same beam,** *when the whole weight is in the centre,* **as 5 to 8.**

Fig. 26.

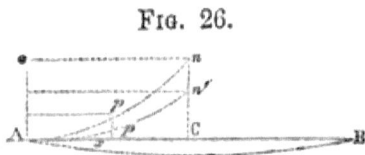

When the whole **weight is at the centre** let w represent the weight upon one of the supports, the strain upon any section at the distance x will be represented by $w\,x$, and the **deflection**, as in the last proposition, by $w\,x^2$. It will, therefore, **as in the last case**, correspond to the abscissa of **a common parabola, of** which x is the ordinate. The sum of these **deflections, or the** whole deflection, will be proportional to **the area $A\,p\,n\,c = \frac{1}{3}$** rectangle $A\,o\,n\,c$.

Let the beam be now supposed to be uniformly loaded, and let the deflection due to the extension of the fibres at the distance x be ascertained. It is evident that the weights upon the points of support will be the same as formerly.

The reaction of the point A may be represented by a force equal to w acting upwards, its leverage at the distance x will be $w\,x$, and, the deflection due to it, $w\,x^2$, as before; but the effect of the uniformly distributed load upon the part x diminishes this deflection, since it acts in the opposite direction; its effect

---

* Tredgold gives the proportion **in this case** as 4 to 3 (see treatise on cast iron, page 141). To test the question by direct experiment, a flexible strip of wood 7 feet long was suspended **at** the middle. Two uniform chains of the same length were laid upon the top, and the deflection found to be $\frac{3}{4}$ of an inch: one chain was then suspended at each end, and the deflection became $\frac{1}{8}^1$ **of an** inch; but, $4:11::3:8\frac{1}{4}$, a result much nearer the calculated proportion than **was** expected with the apparatus used.

will be $\frac{w}{2l} x^3$, and the whole deflection will therefore be $\left(w x - \frac{w}{2l} x^3\right)$. The expression $\frac{w}{2l} x^3$ is represented by the area $A\, p'\, n'\, c$, which we have already shown to be $\frac{1}{3}$ rectangle. And hence, the deflections will be as $\frac{1}{3} - \frac{1}{8} : \frac{1}{3}$, or as 5 to 8.

STRENGTH OF PARTICULAR SECTIONS.

PROP. 13. *Strength of a triangular section.*

As this case is more curious than useful, we will simply indicate the mode of procedure without entering into its full investigation.

FIG. 27.

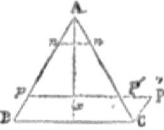

Let $A B C$ represent the section at the point of greatest strain
$h =$ height and $b =$ base of triangle, $p\, p'$ neutral axis
$R =$ the maximum strain upon a superficial unit.

The strain varying as the distance from the neutral axis, it will be $= R$ at the point A, and at any point of $B C$ it will be found thus, $(h - x) : x :: R : \frac{R x}{h - x}$. The strain upon the upper part of the section will be represented by a pyramid, whose base is $A\, p\, p'$, and altitude R; upon the lower part, it will be the wedge-formed solid, whose base is $p\, p'\, B\, C$, and altitude $\frac{R x}{h - x}$. The volume of the solid will be the difference between the wedge $p\, p''\, B\, C$ and pyramid $p'\, p''\, C$.

By equating the moments we obtain the value of x, and, consequently, the position of the neutral axis; this value substituted in the expression for the resistance of the section will give its value in terms of b and h.

It is found, that the **strength of the triangular prism, is to** that of a rectangular prism **having the same base and altitude** as 339 : 1000, or nearly as 1 : 3.

As the resistance to compression and extension are supposed equal, the prism must be equally strong, **whether the** base or **vertex be** compressed, provided the limit of elasticity **be not exceeded.**

The investigation **of** this case **leads** to an apparently paradoxical result: it is found, that **the prism becomes one** thirty-seventh **part stronger** when the **angle** is taken off to one-tenth **of the depth.**

The difficulty will vanish, **when it is remembered that** the **greatest resistance of any** fibre **is** R, **and that in the** triangular **section, only the single point** at the apex **opposes** this resistance; whereas, if a **portion be removed, every point of the** line $n\,n'$ opposes a resistance $= R$.

The triangular section contains **half the surface of the circumscribing** rectangle, but is only **one-third as strong; hence,** there is no **economy in its use.**

PROP. 14. *The **strength of** a cylinder supported at the ends is to **that of its** circumscribed prism as* $\dfrac{3\pi}{16}$: 1, *or as* ·589 : 1.

FIG. 28.

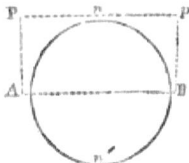

The neutral axis **being at the centre,** the resistances to compression and **extension will be** represented by the ungulas, whose bases are the **semicircles** $A\,n\,B$ and $A\,n'\,B$.

The **volume of the ungula has been found to be** $2\,r^2\,\dfrac{R}{3}$.

The distance of the perpendicular **through the centre of gravity** is $\tfrac{3}{16}\pi r$.*

---

* To find the volume and the position of the centre of gravity of an ungula or solid formed by passing an oblique plane through the diameter of the base of a semi-cylinder.

Figs. 30 and 31.

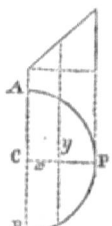

All the sections parallel to AB will be rectangles, the altitudes of which will be proportional to their distance from C.

Hence, if R represent the altitude at p, and $x =$ the distance of any section, $r =$ radius, $y =$ ordinate, $r : x :: R : \dfrac{Rx}{r} =$ perpendicular of rectangle, and $2\dfrac{R}{r} x y =$ area.

The elementary solid will be $2\dfrac{R}{r} x y \, dx$, and its moment $= 2\dfrac{R}{r} x^2 y \, dx$.

The distance to the centre of gravity will be $\dfrac{\int x^2 y \, dx}{\int x \, y \, dx}$.

1. To determine the volume of the ungula we have $y = \sqrt{r^2 - x^2}$; hence, $\int x y \, dx = \int (r^2 - x^2)^{\frac{1}{2}} x \, dx$. Make $r^2 - x^2 = z^2$. Whence, $x \, dx = -z \, dz$, $\int (r^2 - x^2)^{\frac{1}{2}} x \, dx = -\int z^2 \, dz = -\dfrac{z^3}{3} = \dfrac{-(r^2 - x^2)^{\frac{3}{2}}}{3}$,

which becomes, when taken between the limits, o and r, $= \dfrac{-r^3}{3}$.

This **negative** result does not effect the absolute volume: to interpret it, it **must** be observed that the integral does not become o when $x = o$, but when $x = r$, and consequently the solid lying in the direction of C from p should be negative. $\dfrac{2R}{r} \cdot \dfrac{r^3}{3} = \dfrac{2Rr^2}{3} =$ volume required by **substitution in the expression** $\displaystyle\int \dfrac{2R}{r} x y \, dx$.

Hence, the moment becomes $2 r^4 \cdot \dfrac{R}{3} \cdot \tfrac{3}{10} \pi r = \dfrac{\pi r^3 R}{8}$.

The moment of the rectangle $A p$ is $2 r^2 \cdot \dfrac{R}{2} \cdot \tfrac{2}{3} r = \tfrac{2}{3} R r^3$.

But $\tfrac{2}{3} R r^3 : \tfrac{1}{8} \pi R r^3 :: 1 : \dfrac{3 \pi}{16} :: 1 : 589$.

The volume of the ungula is therefore equal to that of a pyramid whose base is the circumscribing rectangle with the same altitude.

To find the centre of gravity, we have $\int x^2 y\, dz = \int (r^2 - x^2)^{\frac{1}{2}} x^2\, dx = \int \dfrac{x}{2} (r^2 - x^2)^{\frac{1}{2}} 2 x\, dx$. Integrate by parts, making in the formula

$$\int z\, dy = z y - \int y\, dz \qquad z = \dfrac{x}{3} \qquad y = (r^2 - x^2)^{\frac{3}{2}}$$

whence, $d y = \tfrac{3}{2} (r^2 - x^2) 2 z\, dz$. Substitute these values we obtain $\int \dfrac{x}{2} (r^2 - x^2)^{\frac{1}{2}} 2 x\, d x = - (r^2 - x^2)^{\frac{3}{2}} \dfrac{x}{3} + \tfrac{1}{3} \int (r^2 - x^2)^{\frac{3}{2}} d x$. The quantity $(r^2 - x^2)^{\frac{3}{2}} \dfrac{x}{3}$ will reduce to o when $x = o$ or $x = r$, this term will therefore disappear and the expression reduces to $\int x^2 (r^2 - x^2)^{\frac{1}{2}} d x = \tfrac{1}{3} \int (r^2 - x^2)^{\frac{3}{2}} d x = \tfrac{1}{3} \int (r^2 - x^2)^{\frac{1}{2}} (r^2 - x^2) d x = \tfrac{1}{3} \int (r^2 - x^2)^{\frac{1}{2}} r^2 d x - \tfrac{1}{3} \int (r^2 - x^2)^{\frac{1}{2}} x^2 d x$. Transpose the last term to the first number, and reduce $\int (r^2 - x^2)^{\frac{1}{2}} x^2 d x = \tfrac{1}{4} (r^2 - x^2)^{\frac{1}{2}} r^2 d x$.

But the integral $\int (r^2 - x^2)^{\frac{1}{2}} (dx)$, between the limits o and r, represents the area of a quadrant $= \tfrac{1}{4} \pi r^2$.

Hence, $\int (r^2 - x^2)^{\frac{1}{2}} x^2 d x = \tfrac{1}{4} r^2 \left(\dfrac{\pi r^2}{4} \right) = \dfrac{\pi r^4}{16}$ and $\dfrac{\int (r^2 - x^2)^{\frac{1}{2}} x^2 d x}{\int (r^2 - x^2)^{\frac{1}{2}} x^2 d x}$

$= \dfrac{\dfrac{\pi r^4}{16}}{\dfrac{r^3}{3}} = \tfrac{3}{16} \pi r =$ distance from the centre of the circle to the perpendicular through the centre of gravity.

The line which joins C and the middle point of R passes through the centres of all the elementary rectangles, and, therefore, the centre of gravity must be found at the intersection of this line with the perpendicular, through a point at a distance of $\tfrac{3}{16} \pi r$ from the centre. Its distance from the base is therefore the fourth term of the proportion $r : \tfrac{3}{16} \pi r :: R : \tfrac{3}{16} \pi R$.

54   BRIDGE CONSTRUCTION.

Prop. 15. *If n represent the ratio of the inner and outer diameters, the strength of the solid cylinder will be to that of a tube of the same exterior diameter as* $1 : (1 - n^4)$.

Fig. 29.

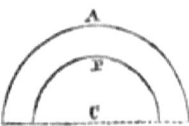

Let r = exterior radius, $n\,r$ = interior radius, R = strain at A, then $r : n\,r :: R : n\,R$ = strain at P.

The resistance of the semicircle CA is $\dfrac{\pi r^3 R}{8}$. (See last problem).

The resistance of the semicircle CP is $\dfrac{\pi n^4 r^3 R}{8}$.

The resistance of the ring is $\dfrac{\pi r^3 R}{8}(1 - n^4)$

$$\dfrac{\pi r^3 R}{8} : \dfrac{\pi r^3 R}{8}(1 - n^4) :: 1 : (1 - n^4).$$

Prop. 16. *To find the strength of a vertical rib with horizontal flanges on both sides.*

Fig. 32.

This case can be immediately deduced from that of a rectangular section, for the area is evidently equal to the rectangle $AC -$ 2 rectangles $n\,m$, and as the neutral axis is in the centre, the strength will be equal to the difference of the strength of these rectangles. Call $b\,d$ = the circumscribing rectangle, $A\,C,\,b'\,d'$ the deducted rectangles $2\,n\,m$.

But the strain on the extreme fibres of the inner rectangle is not as great as R; it is therefore necessary to introduce the relative values of these strains. The strain at the distance $\tfrac{1}{2}\,d'$ from the neutral axis is determined from the proposition

$$d : d' :: R : \frac{R\,d'}{d}$$

$b\,d^2\,R$ represents the resistance of the rectangle $b'\,d'$

$b'\,d'^2\,\dfrac{R\,d'}{d}$ = resistance of the rectangle $b'\,d'$

These resistances are to each other as $b\,d^2 : \dfrac{b'\,d'^3}{d} :: b\,d^3$: $b'\,d'^3$.

Hence, the strength of the circumscribed rectangle is to that of the given section as $b\,d^3 : (b\,d^3 - b'\,d'^3)$.

PROP. 17. *The strongest beam that can be cut out of a tree or given cylinder has the breadth and depth in the proportion of 1 to $\sqrt{2}$.*

FIG. 33.

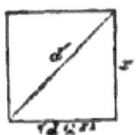

The strength is as $b\,d^2$, or as the product of the breadth and square of the depth. Call d' = diameter of cylinder, x = depth; then, $\sqrt{d'^2 - x^2}$ = breadth, and $x^2\sqrt{d'^2 - x^2}$ = a maximum. The first differential co-efficient is $4\,d'^2 x^3 - 6\,x^5$, placing this equal to zero, x becomes $= d'\sqrt{\tfrac{2}{3}}$; substitute this value of x in the expression for the breadth, it becomes

$$\sqrt{d'^2 - \tfrac{2}{3}\,d'^2} = d'\sqrt{\tfrac{1}{3}}.$$

But $d'\sqrt{\tfrac{2}{3}} : d'\sqrt{\tfrac{1}{3}} :: \sqrt{2} : 1$.

In precisely the same way, it can be shown, that the stiffest beam which is proportional to $b\,d^3$ will have its breadth to its depth as 1 : √3. In this case the **breadth** is equal to the radius.

The geometrical construction of the figure of the stiffest beam is extremely simple. From opposite extremities of any diameter with radii equal to the radius of the cylinder describe arcs cutting the circumference and join the points of intersection.

FIG. 34.

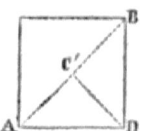

The construction of the figure of the strongest beam is also very simple, for, since the sides are as 1 : √2, the hypothenuse will be √3, and from the properties of right-angled triangles,

$\sqrt{3} : 1 :: 1 : AC = \dfrac{1}{\sqrt{3}}$; but, $\dfrac{1}{\sqrt{3}} : \sqrt{3} :: 1 : 3$; hence, $AC = \tfrac{1}{3} AB$.

Lay off therefore one third of the diameter, and erect a perpendicular; its intersection with the circumference will determine the point D.

PROP. 18. *To find the resistance of a beam lying horizontally upon an edge.*

FIG. 35.

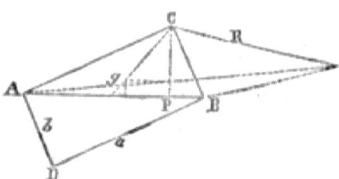

Let AB be a diagonal h = perpendicular CP R = strain upon C, the whole strain will be represented by the pyramid, whose base is ABC and altitude R; its volume is

$\frac{bd}{2} \cdot \frac{R}{3} = \frac{bdR}{6}$. The distance from AB to the perpendicular through the centre of gravity is $\frac{1}{2}h$, and the moment will therefore be $\frac{bdhR}{12}$. The moments of the two pyramids ABC and ABD will be $\frac{bdhR}{6}$

When the side AC is vertical the moment is $\frac{bd^2R}{6}$.

The ratio of the strength will therefore be as $d : h$, or as the side is to the perpendicular.

When the section becomes a square we have $d : h : : \sqrt{2} : 1$; hence, the strength of a square beam where the side is vertical is to the strength when the diagonal is vertical as $\sqrt{2} : 1$.

PROP. 19. *When the pressure upon a beam supported at the ends varies as the distance, the point of greatest strain will be at a distance from the unloaded extremity equal to the length multiplied by the square root of one-third, or* $(l \sqrt{\frac{1}{3}})$.

FIG. 36.

In this case, the pressure will be proportional to the area of a triangle, the centre of gravity of which is at a distance from $A = \frac{2}{3}l$.

Then, for the weight upon A we have $l : \frac{l}{3} : : w : \frac{w}{3} =$ weight at A.

In like manner $\frac{2w}{3} =$ weight at B.

Let $x =$ distance of any section from A. The resistance of the support A, being regarded as a force $= \frac{w}{3}$ acting upwards, will be **opposed** by the weight upon x acting downward, and the difference **of** the moments **will represent the strain.**

For the first we have $\frac{w}{3} \cdot x = \frac{wx}{3} =$ moment of force $\frac{1}{3} w$.

For the second we have, since the **weights will** be as the square of the lengths, $l^2 : x^2 :: w : \frac{wx^2}{l^2} =$ weight on x. As the leverage is $\frac{x}{3}$ $\frac{wx^2}{l^2} \cdot \frac{1}{3} x = \frac{wx}{3 l^2} =$ moment of the weight on x.

The difference will be $\left(\frac{wx}{3} - \frac{wx^3}{3 l^2}\right) =$ strain on section.

By the principles of maxima **and** minina we have

$$\left(\frac{w}{3} - \frac{w}{l^2} x^2\right) = o. \quad x^2 = \frac{l^2}{3}. \quad x = l\sqrt{\tfrac{1}{3}}.*$$

PROP. 20. *To determine the extension of the fibres when a beam is supported at the ends and loaded in the middle.*

A beam supported at the ends and loaded in the middle **is** in the **same** condition as **a** beam resting upon **a** fulcrum in the **middle and** loaded with equal weights at the ends.

FIG. 37.

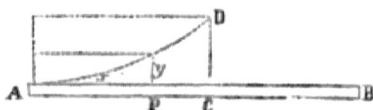

Let $l =$ one-half the **whole length**
$w =$ the weight on A
$e =$ the maximum extension, **which will be at** C.

Now, as the extension **at any distance** is **in** proportion to the strain, it will evidently be in proportion to x; and we have therefore, $l : x :: e : \frac{ex}{l} =$ extension at the distance x.

---

* Tredgold gives it $\sqrt{\tfrac{1}{3}}\, l$.

But, the deflection being as the extension and distance directly, and inversely as the depth, it will be as $\dfrac{e\,x}{l} \cdot \dfrac{x}{d} = \dfrac{e\,x^2}{l\,d}$.

Call this expression (y), we have therefore, $y = \dfrac{e}{l\,d} x^2 =$ the equation of a parabola, of which, x is the ordinate, and y the abscissa.

The whole deflection being equal to the sum of these abscissas will be represented by the area $A\,C\,D = \frac{1}{3}$ rectangle $A\,D = \frac{1}{3} l \cdot \left(\dfrac{e}{l\,d} l^2\right) = \dfrac{e\,l^2}{3\,d}$. The deflection of the part $B\,C$ being equal to that of $A\,C$, the whole deflection will be $\dfrac{2\,e\,l^2}{3\,d}$.

Whence, $\dfrac{3\,d \times \text{(deflection)}}{2\,l^2} = e.$*

By observing the deflection produced by a given weight, and substituting its value in the above expression, the value of e can be ascertained. For cast iron, when the weight is 15,300 lbs. per square inch, it is found to be $\frac{1}{1201}$ inches for a length of one inch.

Means of determining the constants.

The equation $R' = \dfrac{B\,D^3}{w\,l^2}$ expresses the relation between the dimensions of a beam when the deflection is in proportion to the length. If this deflection be assumed at $\frac{1}{40}$ of an inch for every foot of length, and $d =$ the observed deflection caused by the application of a weight w.

Then, $d : w :: \dfrac{l}{40} : \dfrac{w\,l}{40\,d} =$ weight required to produce the given deflection. By substituting this value for w, we obtain $R = \dfrac{40\,B\,D^3\,d}{w\,l^3}$.

To determine the value of R, experimentally, let a beam

---

* This expression is the same as that given by Tredgold, but the manner of obtaining it is far more direct and simple. (See Treatise on cast iron, p. 136.)

be placed upon **two supports** and loaded with any known weight, not **so great as to impair** the elasticity.

Observe the **deflection (d), the** weight (w), the distance be-**tween the** supports (l), the **breadth** (B), and the depth (D). **The** length being in feet, and the other dimensions in inches.

Substitute these **values and** perform the operations indica-**ted, the value of** R will **be obtained.**

This constant for cast iron has been found to be ·001 Tredgold.
" " White fir " " " ·01 "
" " Oak " " " ·0109 "
" " Yellow pine " " ·0115 "
" " American white pine " ·0125 Author.

The formula which expresses the strength of a beam is $\dfrac{3\,w\,l}{2\,b\,d^2} = R$ **when the beam** is supported at both ends and the weight applied in the middle.

To determine the constant, weights should be applied **and** gradually increased as long as no perceptible flexure remains upon their removal.

The highest **value of w thus** obtained will give the value of R. In this **formula,** R expresses the maximum strain **upon a square inch; but,** in determining its value, when **used in proportioning the parts of** important structures, it **is proper to observe, that** the strength of materials generally dimin-ishes **as** the length **of time** in use increases, and, that a weight which will produce **no** perceptible deflection in a short time, may produce a very great deflection when **long** continued.

From some experiments, made by the writer **in** the spring of 1840, it appeared, **that** locust would bear for **a** few seconds a strain of 5500 pounds per square inch without apparent injury, but the elasticity was impaired by 2304 pounds per square inch continued 16 days.

The value of R for cast iron **when the** time was short was **found by** Tredgold to be 15300 pounds
    For White fir the constant is 3519 "
    " Oak " " " 3825 "
    " Yellow pine " " 3825 "

The **above constants were** deduced from experiments and

other data furnished by Tredgold in his treatise on **cast iron**, but the writer believes them to be entirely too great for permanent strains: in the course of experiments, already referred to,* he **found, that** when white pine, **yellow pine, and** hemlock, were subjected to a strain of from 1500 lbs. to 1800 lbs. per square inch, continued for 16 days, the pieces did not recover

* These experiments were made at York in the year 1840. The writer upon commencing his duties as an engineer on the York and Wrightsville Rail Road, found in the office a number of very fine specimens of wood, that had been procured, for the purpose of experimenting upon them, by T. Jefferson Cram, formerly, assistant professor of natural philosophy at the U. S. Mil. Acad., and at that time, a civil engineer on the Baltimore and Susquehanna Rail Road. Unwilling to lose so favorable an opportunity, several other specimens were added to the number, and experiments made with **a view of** determining, not the absolute strength, but **the** elastic limit. The result is given in the following table:

When the pieces did **not** exactly recover their shape they are marked injured.

	Kind of Timber.	Strain per sq. inch.	Time.	Remarks.
1	White pine	2272	10 min.	Injured.
2	do.	1548	16 days	Injured.
3	Hemlock	2624	5 min.	Injured.
4	do.	1620	16 days	Injured.
5	Yellow pine	2848	5 min.	Injured.
6	do.	1800	16 days	Injured.
7	Locust	5504	2 min.	Not injured.
8	do.	3600	$3\frac{1}{2}$ days	Injured.
9	do.	2304	16 do.	Injured.
10	White oak	4248	15 min.	Not injured.
11	do.	7200	do.	Injured.
12	do.	3648	40 hours	Not injured.
13	do.	4088	48 **do.**	Injured.

The pieces were all 5 feet long, 3 inches deep, 1 inch wide, supported at one foot from the end. The weight acting with a leverage of 4 feet. R is determined from the formula $R = \dfrac{6\,w\,l}{b\,d^2}$.

There were three pieces of each kind, all very superior.

From these experiments it appears, that there is a great difference in the powers of resistance of different kinds of timber, and that Oak and Locust are far superior to Hemlock and Pine. Also, that a small weight long continued may produce more permanent flexure than a much greater one applied only for an instant.

their shape upon the removal of the weight; and in practice, he has not considered it safe to assign more than 800 lbs. per square inch as a permanent load and 1000 as an accidental load.

The following table may be found of much utility:

Column A contains the constants used in the formula for the stiffness of beams supported at the ends and loaded in the middle. $R = \dfrac{B D^3}{w\, l^2}$, (dimensions all in inches, except the length, which is in feet.)

Column B contains the constants used in the formula for the strength $R = \dfrac{3\, w\, l}{2\, b\, d^2} =$ strain on a square inch in pounds, (the dimensions being all in inches.)

Column C gives the greatest extension without injury.
    "   D   "   specific gravity.
    "   E   "   weight of a cubic foot in lbs.
    "   F   "   wt. of the modulus of elasticity in lbs.
    "   G   "   height of the mod. of elasticity in feet.

The data have been obtained from Tredgold.

	A	B*	C	D	E	F	G
Cast iron	·001	15,300	$\frac{1}{1204}$	7·2	450	18,400,000	5,750,000
Malleable iron	·0008	17,800	$\frac{1}{1400}$	7·2	475	24,920,000	7,550,000
White fir	·01†	3,630	$\frac{1}{384}$	0·47	29·3	1,830,000	8,970,000
Oak	·0109	3,960	$\frac{1}{230}$	0·83	52	1,700,000	4,730,000
Yellow pine	·0105	3,900	$\frac{1}{214}$	0·46	26¾	1,600,000	8,700,000

The weight of the modulus of elasticity is determined by the proportion $1 : \dfrac{1}{e} :: w :$ modulus. (e) is found in column C, and (w) in column B.

---

* The constants in column B I consider too great for timber; as regards the others I can express no opinion, having made no experiments. As a general rule, I should not think it safe in practice to use higher numbers than those given in column B divided by four.—*Author*.

† By experiments of Author on American white pine the constant is ·0125.

☞ In applying the above formulas it must be observed, that w includes the weight of the beam itself. To find the weight sustained at the middle point, one-half of the weight of the beam must be deducted; or, if the load is uniformly distributed, deduct the whole weight of the beam and multiply by 2.

WOODEN BRIDGES.

INSTEAD of commencing this treatise with a history of Bridge Construction, and an explanation of the various plans that have from time to time been adopted, it is believed that a preferable mode will be to establish first the true principles of construction, and then to apply these principles to an examination of plans that have been executed. By this means it can be ascertained how far they coincide with, or depart from, the principles which we endeavor to establish, and how far the correctness of these principles is confirmed by practical experience.

The most important part of any bridge, and that which admits of the greatest variation in form and principle, is the support of the roadway. To this therefore our chief attention will be directed.

The most simple support for a roadway evidently consists of a series of longitudinal timbers laid between two abutments or piers. And as the examination of this case will lead, by easy gradations, to others which are more complicated, and as it also involves many of the principles which apply to structures of a more important character, we will begin by an examination of the forces which act upon a single beam laid upon two supports and loaded with a weight, either uniformly distributed, or concentrated at any given point.

It has been shown in treating of the resistance of solids, that the fibres on the upper side will be compressed, and on the lower side extended; that within the elastic limits the resistances to these forces are equal; that the intensity of the strain varies directly as the distance of any fibre from the neutral axis, and that at the axis itself the strain is nothing.

There exists, also, a force called, by Dr. Young, detrusion, the effect of which is to crush across the fibres close to a fixed point, and the resistance to which is directly proportional to the area of the cross section.

This force, as has been shown, modifies the form of a beam of equal strength, which, instead of being the apex of a conic section at its extremity, must be enlarged sufficiently to resist this force of detrusion.

The existence of this vertical force, and its effects at other points, have not been considered by writers on the resistance of solids, probably because it diminishes rapidly in approaching the centre of a beam, whilst the area of the section generally increases. That a vertical strain upon the fibres exists at other points can be shown by the following considerations.

Let AB represent a beam supported at A and B, and disregarding for the present its own weight, let it be loaded with a weight applied at the centre.

Fig. 38.

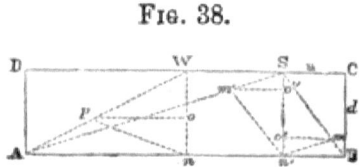

This force is directly transmitted to the points A and B, each of which sustains one half the weight. The lines of direction of the forces are along AW and BW.

By constructing the parallelograms of forces on the diagonals, we find $wo = \frac{1}{2} w$ $n = \frac{1}{2} w$ for the vertical forces transmitted to A and B, and $po =$ the horizontal strain at w, which is determined by the proportion $d : \frac{1}{2} l :: \frac{1}{2} w : op = \dfrac{wl}{4d}$, (in which l represents the length, and d the depth.) The same force is transmitted to B. We can also determine this horizontal force, by the condition that it shall keep the part $w\,c$ in equilibrio. Regarding w as a fulcrum, and the weight at $B = \tfrac{1}{2} w$, the moment of this force will be $\tfrac{1}{2} w \times \tfrac{1}{2} l = \dfrac{w\,l}{4}$. The moment of the horizontal force, acting with a leverage d, will be $H d$, and $H = \dfrac{w\,l}{4\,d}$ as before.

We will now consider the action of these forces at another point (S), the weight, as before, being applied entirely at the middle point of the beam.

1. Horizontal strain at S.

Since the weight w is equally supported by each of the points A and B, we may continue to consider (w) as a fulcrum, and, that forces ($\tfrac{1}{2} w$), acting upwards at A and B, maintain the equilibrium.

The portions ($A\,n$) and ($n\,B$) will be in the condition of beams fixed at one end, and loaded at the other.

The weight $\tfrac{1}{2} w$ applied at B, acting with a leverage $u = S\,C$ produces an effect equal to the product $\dfrac{w}{2} \times u$, and the horizontal strain at S acting with a leverage d has for its moment $H \times d$. Hence, $H d = \dfrac{w\,u}{2}$ or $H = \dfrac{u\,w}{2\,d}$, which becomes when ($u = \dfrac{l}{2}$), $w = \dfrac{l\,w}{4\,d}$, as before.

The horizontal strain in the middle of the beam is to the same strain at any other point as $\dfrac{l}{2} : u$, and consequently varies with the perpendiculars of a triangle constructed on $\dfrac{l}{2}$ as a base.

2. Vertical force at any point.

The horizontal at S was found to be $\dfrac{u\,w}{2\,d}$, but it is evident, that the portion of the beam DS, with the applied weight at W, presses against the cross section at S, and must be resisted by the reaction at that point.

If the horizontal force $\frac{uw}{2d}$, acting with its leverage (d), was sufficient to sustain the part DS, the effect of the weight at W would be entirely overcome, and there would remain nothing to produce a downward strain upon the fibres at (S), or, in other words, the vertical force would be zero. That this is not the case, however, can be seen by estimating the force necessary to sustain DS in equilibrium.

As (W) acts with a leverage DW or An, the equation of moments will be $H'd = \frac{wl}{2}$, or $H' = \frac{wl}{2d}$. But we have seen that the horizontal strain at S is actually $H = \frac{uw}{2d}$.

The difference is $(H' - H) = \frac{w}{2d}(l - u)$.

As this expression cannot become zero for any point between W and C, it follows, that the horizontal force is not sufficient to sustain the weight, and there must consequently be a cross strain upon the fibres which must compensate for this deficiency, and be resisted by a vertical reaction.

Call this vertical force f, it acts with a leverage $= DS = (l - u)$. The difference of the horizontal forces, or $H' - H$, acts with a leverage $= d$. The equation of moments will therefore be $f(l - u) = d \cdot \frac{w}{2d}(l - u)$, from which, we obtain $f = \frac{w}{2}$, or, *the cross strain upon the fibres produced by a weight applied in the middle is constant at every point, and equal to one-half the weight.*

We can obtain the same result by another method. Using the same notation as before, we may suppose that the vertical force (f) [acting at S with a leverage $SD = (l - u)$,] and the horizontal strain at S, acting with a leverage d, sustain in equilibrium the weight (W) acting with a leverage $WD = \frac{l}{2}$.

WOODEN BRIDGES. 67

The equation of moments will be

$$H d + f(l - u) = \frac{w l}{2}.$$

Substitute the value of $H = \dfrac{u w}{2 d}$ and reduce we find $f = \dfrac{w}{2}$ as before.

Again, let (W) be a fulcrum and ($\frac{1}{4} w$) a force acting upwards at B, with a leverage $W C = \dfrac{l}{2}$, W being taken as the point of rotation. This force will be resisted by the horizontal force acting with a leverage (d), and f acting with a leverage $(\dfrac{l}{2} - u) = W S$.

The equation of moments will be

$$H d + f \left(\frac{l}{2} - u\right) = \frac{w l}{4},$$

which, by reduction, gives $f = \dfrac{w}{2}$, as in the other cases.

When the weight is not applied in the middle, it may be shown in the same way, that the vertical forces on each side will be constant and equal to the pressures on the points of support.

Let the weight be uniformly distributed.

In this case the forces will be determined in two different ways, as this will serve to verify the results, and exhibit more fully the manner of their action and distribution.

1. By means of the moments.

The weight being uniformly distributed, one-half of it will be sustained by each point of support. To estimate the strain at any point, S, we may suppose the part $D S$ to be fixed, and the weight $\frac{1}{2} w$ at B, to act upwards with a leverage u, this force is opposed by the weight of the portion u acting downwards with a leverage $\dfrac{u}{2}$; hence, if $H' =$ strain at S, $H' d =$

$\dfrac{w u}{2} - \dfrac{u^2 w}{2 l}$, $H' = \dfrac{u w}{2 d}, \left(\dfrac{l - w}{l}\right) =$ horizontal strain at S,

when $u = \dfrac{l}{2}$ we have $H = \dfrac{w l}{8 d} =$ horizontal strain at centre

2. By resolution of forces.

The weight at S is one-half of the portion on SD and one-half of that on SC, consequently, it is equal to one-half of the whole weight, or $\frac{w}{2}$. Join SA and SB, and let $Sn' = \frac{1}{2} w$. From similar triangles we can obtain

$$AB : An' :: Sn' : So'$$
$$\text{or} \quad l : l - u :: \frac{w}{2} : So' = \frac{w(l-u)}{2l}$$

= weight transmitted to B from S,

and $\frac{w}{2l}(l-u) + (\frac{w}{2l}) u = \frac{1}{2}$ the weight on $u) = \frac{w}{2}$, as it should be, for the whole weight at B.

We also have $Sn' : n'B :: So' : o'm = \dfrac{So' \times n'B}{Sn'} = \dfrac{w(l-u)}{2ld} \times u = \dfrac{uw}{2d}\left(\dfrac{l-u}{l}\right)$ for the horizontal force at S, as before.

The same results would be attained, for $m'o''$ = the horizontal force in the direction Sd.

The horizontal force that would sustain the part SC, as found by taking the moments, is $\dfrac{uw}{l} \times \dfrac{u}{2} \times \dfrac{1}{d} = \dfrac{uw}{2d} \cdot \dfrac{u}{l}$.

Comparing this with the former result, we find that the horizontal strain at S is to the force that would simply sustain the part SC as $(l-u)$ is to u; and consequently, it is only when $u = \dfrac{l}{2}$ or the point S coincides with the centre, that the horizontal strain is equal to the force that would sustain either portion in equilibrio, if the beam be supposed to be cut through at that point.

As $(l-u)$ is greater than u for any point between C and the centre, it follows, that the horizontal strain is greater than would be produced simply by the pressure of the weight on SC if free to move around B.

If we make $u = o$, both the above strains become o, but the proportion $l - u : u$ would give $l : o$. This result is a consequence of the omission of the factor u which is common to both terms; if it be introduced, the proportion becomes $u \,(l-u) : u^2$, which reduces to $o : o$ when $u = o$ and involves no contradiction.

The strain at any point being $\frac{w}{2\,d}\left(u - \frac{u^2}{l}\right)$ will be a maximum whenever $\left(u - \frac{u^2}{l}\right)$ is a maximum. The first differential coefficient of this expression being placed $= o$, we have $1 - \frac{2\,u}{l} = o$, or $u = \frac{l}{2}$. Hence, the maximum strain is in the centre.

The force necessary to sustain the part SD, if applied at S, in the direction SD, is $\frac{w}{2\,d}\left(\frac{l-u}{l}\right)^2$.

Comparing this with the actual strain at S, which is $\frac{w}{2\,d}\left(\frac{l-u}{l}\right)u$, we find the former to be the greater in the proportion of $(l-u) : u$.

Consequences which are deduced from the above results.

1. Since the horizontal strain at the centre is exactly equal to that which would sustain one-half of the beam, if the other half should be removed, it follows, that there can be no vertical strain at this point.

2. As the actual horizontal strain at any other point S is less than the force that would sustain the part SD in the proportion of u to $l-u$, it follows, that the part SD must press vertically and produce a vertical strain, which would be measured by a vertical force sufficient to compensate for the difference of the horizontal forces.

As the horizontal forces are proportional to u and $(l-u)$. If H represent the strain at S, we have $u : l - u :: H : H\left(\frac{l-u}{u}\right)$, and the difference of the forces will be $H\left(\frac{l-u}{u}\right) - H = H\left(\frac{l-2\,u}{u}\right)$.

This force is not resisted by any antagonist horizontal force, and must therefore produce a vertical strain on the fibres, the measure of which would be the force which acting vertically would give the same moment in reference to the point A.

Call w' this force; its leverage is $(l-u)$. The moments will be $H\left(\dfrac{l-2u}{u}\right)d$ and $w'(l-u)$, by which we obtain $w' = H d \left(\dfrac{l-2u}{u(l-u)}\right)$. Substitute the value of $H = \dfrac{wu}{2d}\left(\dfrac{l-u}{l}\right)$ we have $w' = \dfrac{w}{2l}(l-2u) =$ The expression for the vertical strain at any point S which becomes zero at the centre, or when $u = \tfrac{1}{2}l$. When $u = o$ we have $w' = \dfrac{w}{2} =$ strain at the ends.

This expression $\dfrac{w}{2l}(l-2u)$ is equal to the difference $So' - So''$ of the vertical components for

$$l : l-u :: \tfrac{1}{2}w : \dfrac{w(l-u)}{2l} = So'$$

$$l : u :: \tfrac{1}{2}w : \dfrac{wu}{2l} = So''$$

$$\dfrac{w(l-u)}{2l} - \dfrac{wu}{2l} = \dfrac{w}{2l}(l-2u)$$

Hence $o'\, o''$ represents the vertical strain at any point.

The expression $\dfrac{w}{2l}(l-2u) = -\dfrac{w}{l}u + \dfrac{w}{2}$ bring of the form. $y = -ax + b$ is the expression for a straight line, and therefore if AB represent a beam, and $Bn = \tfrac{1}{2}w$ the straight line, cn will determine the perpendiculars $p\,m$ which measure the strains.

FIG. 39.

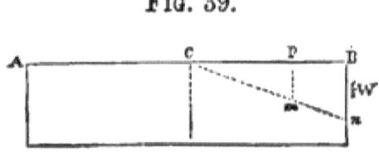

To show this more clearly let $u = \dfrac{l}{2} - x$ whence $x = \dfrac{l}{2}$

$-u = cp$. Substitute the value of u and reduce: the expression $-\dfrac{w}{l} u + \dfrac{w}{2}$ becomes $\dfrac{w}{l} x$ which is the equation of a straight line passing through the origin C.

It follows therefore that the vertical strains are exactly proportional to the distance from the centre, a consequence of the greatest importance in its application to the practice of Bridge Construction.

To find the curve which represents the horizontal strain.

The horizontal strain at any point of a beam supported at the ends and loaded uniformly, was found to be $\dfrac{u\,w}{2\,d} \times \dfrac{(l-u)}{l}$ in which u represents the distance of the point from the end, $w =$ the whole weight, $l =$ the length, and $d =$ the depth of

NOTE.—It may be thought that the principle which I have endeavored to establish is too simple to require the explanation that has been given; but simple as it is, the consequences are important, and I do not know that it has been noticed by writers upon Bridge Construction or the resistance of Solids; certain it is that the effects which naturally result from it have been overlooked in proportioning structures. In fact it was not until some months after my attention had been directed to the theory of Bridge Construction, that I was led to observe the difference in the vertical forces at different points of a straight truss. The fact that such difference exists was first pointed out to me by H. R. Campbell, of Phila., a gentleman who in the course of a long and extensive practice as a Civil Engineer, has enjoyed rare opportunities for becoming acquainted with Bridge Construction and for observing the effects of time and accidents.

In a conversation with him upon the principles of the art, he asked me to explain why the chords of a Bridge which had settled considerably were more bent at the abutments than at the middle. I had not then particularly noticed the fact, but he assured me that although the depression was greatest in the middle when a straight Bridge settled below its level, yet the curvature was not uniform, and the quickest bend, or in other words the least radius of curvature, was always nearest the abutment. In a subsequent examination of a large number of bridges, I invariably found that the joints of the braces near the abutments were compressed and tight, whilst near the centre of the spans no symptoms of crushing were perceptible, and in some cases where the joints of the central braces were not well fitted, a knife blade could be introduced, clearly indicating a great increase of pressure towards the abutments, and as a consequence, the necessity of increasing the number or size of the vertical supports towards the extremities.—*Author.*

the beam. This expression can be put under the form $\frac{w}{2d}u - \frac{w}{2dl}u^2$.

Fig. 40.

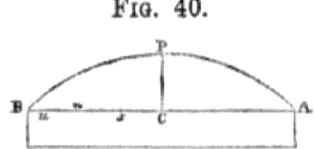

If we make $u = o$ or $u = l$ the expression in either case becomes o, and if we express the values of the strains by the ordinates of a curve of which the above is the equation, we find that the curve passes through the points B and A at the ends of the beam.

If we differentiate the expression $\frac{w}{2d}u - \frac{w}{2dl}u^2$ and place the first differential co-efficient equal to zero: we have $\frac{w}{2d} - \frac{w}{2dl}u = o$ whence $u = \frac{l}{2}$ and the maximum is at the centre C.

The value of the maximum strain found by substituting $\frac{l}{2}$ for u is $\frac{wl}{8d}$. Let this value be represented by the line C, p and p will be a point of the curve.

To ascertain the nature of the curve, we will transfer the origin from B (the point from which u is reckoned) to p.

First make $u = (\frac{l}{2} - x)$ we obtain for the value of the ordinate when the expression is reduced $y = \frac{wl}{8d} - \frac{w}{2dl}x^2$ which is the equation of the curve when the origin is at C.

Again make $y = \frac{wl}{8d} - y'$ and we obtain $y' = \frac{w}{2dl}x^2$, which is the equation of the curve when the origin is at p, but this equation is that of a parabola; hence, the ordinate of a parabola drawn through B, p, and A will exhibit the intensity of the horizontal strain at any point, and furnishes a geometrical method of obtaining it.

WOODEN BRIDGES.

To find the pressure upon the supports when a beam is framed as a cap upon the tops of several vertical posts, and a weight applied directly over one of the posts.

This is a case which may be of use in proportioning the timbers for bridges when the supports are close together.

Fig. 41.

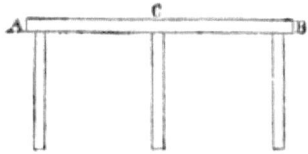

If we suppose the material of the posts to be perfectly inelastic, the middle one would bear the whole of a weight applied at C and no part of it would be sustained by A and B: but if the beam be flexible and the substance of the posts elastic, the pressures upon A and B would depend upon the relative degrees to which these properties were possessed. If the beam be very stiff and the posts elastic, a large part of the pressure will be thrown upon A and B, and if the beam be very flexible and the post but slightly elastic, nearly all the weight will be sustained at C.

When the distance between the supports and the dimensions of a beam are known, the flexure caused by a given weight is readily calculated: and when the length of a support is known, the reduction in length due to a given weight can also be determined.

If w represent the weight at C, $d =$ the deflection which would be produced if the support were removed, $e =$ the reduction in length by the same weight which the post would experience. Then if x represent the actual deflection, we will have, since the deflection is always proportional to the weight,

$$d : x :: w : \frac{w\,x}{d} = \text{weight}$$ sustained by the beam and which is transmitted to the points A and B.

Also, $e : x :: w : \dfrac{w\,x}{e} =$ weight sustained by post C, the sum of these weights must be equal to w. We therefore have

$$\frac{w\,x}{d} + \frac{w\,x}{e} = w \text{ or } (e+d)x = e\,d, x = \frac{e\,d}{e+d}$$

Substituting this value we find

For the pressure upon A and $B = \dfrac{w\,x}{2\,d} = \dfrac{w\,e}{2\,(l+d)}$

For the pressure upon $C = \dfrac{w\,x}{e} = \dfrac{w\,d}{l+d}$

If the ends of the posts instead of resting against solid points of support, be placed upon a second beam, the circumstances of the case will be very different.

Fig. 42.

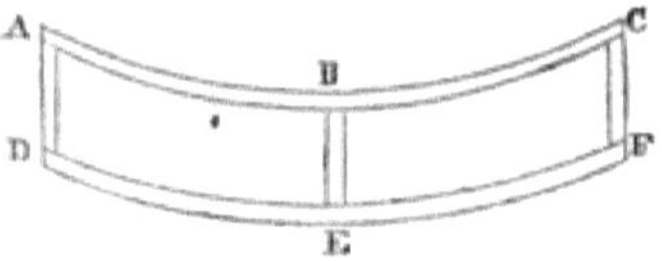

Let $A\,C$ and $D\,F$ be two equal beams connected by an upright in the centre and loaded with a weight at B.

If we suppose $B\,E$ to be perfectly incompressible, then in case of flexure $a\,c$ and $D\,F$ would retain their parallel positions, and each would assist equally in sustaining the load, the post would then be pressed upwards against the point B with a force equal to the reaction of the lower beam or equal to $\dfrac{w}{2}$.

But if the post be elastic it will be compressed to some extent by the action of $\dfrac{w}{2}$, and as a consequence, $D\,F$ would rise, and the deflection becoming less it would sustain less of the weight. $A\,C$ must then sink lower to compensate for this diminished strain on the lower beam, and in proportion to the elasticity of $B\,e$ will be the difference of the strains upon $A\,c$ and $D\,F$.

To determine the strains and deflections of the beams and the degree of compression of the posts by calculation. Let the beams be supposed of any relative size, and to make the case general, let the stiffness of the lower be to that of the upper as $n : 1$.

Also let $w =$ weight at B, $d =$ the deflection that it would produce in the distance $A\ c$, $e =$ the compression of the post by the same weight, $x =$ the actual deflection in the upper beam. Then $d : x :: w : \frac{w}{d}\ x =$ weight sustained by upper beam, and $w - \frac{w}{d}\ x = w\left(1 - \frac{x}{d}\right) =$ portion of weight transmitted to lower beam.

The deflection of the lower beam by the weight w is $n\ d$, hence the actual deflection will be determined by the proportion $w : n\ d :: w\left(1 - \frac{x}{d}\right) : n\ d\left(1 - \frac{x}{d}\right) = n\ d - n\ x = n\ (d - x)$.

The difference between the deflections of the beams must give the compression of the post, which is accordingly equal to $x - n\ (d - x) = (n + 1)\ x - n\ d$. But the compression of the post as determined from the pressure will be $w : e :: w\left(1 - \frac{x}{d}\right) : e\left(1 - \frac{x}{d}\right)$, equating these results we have $e - \frac{e\ x}{d} = (n + 1)\ x - n\ d$, whence $x = \frac{d\ (n\ d + e)}{(n + 1)\ d + e}$. This value substituted in the expressions $\frac{w}{d}\ x$ and $w\left(1 - \frac{x}{d}\right)$ will give the portions of the weight sustained by the upper and lower beams, and by the post.

From the above we learn that when the beams are equal, the pressure upon the post is always less than $\frac{1}{2}\ W$.

For in this case $n = 1$ and $x = \frac{e\ d + d^2}{2\ d + e}$ when $e = o$ or the post is incompressible, we have $x = \frac{d^2}{2\ d} = \frac{d}{2}$, and each beam bears half the weight, consequently the strain upon the post, which is always equal to that upon the lower beam, will be $\frac{1}{2}\ W$. If e be not o the value of x will be greater than $\frac{d}{2}$, and consequently the post will transmit to the lower beam less than $\frac{w}{2}$.

The pressure upon the points A and C will be each one half of the weight sustained by the upper beam, and on the points D and F one-half of the weight on the lower beam.

Strength of a long beam laid over several supports.

This subject properly belongs to the resistance of timber; but as it expresses so nearly the condition of a continuous bridge supported by a number of piers, it has been considered preferable to introduce it in this place.

FIG. 43.

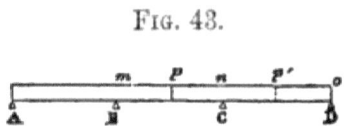

Let $A\ B\ C\ D$ represent a beam laid over several supports, and loaded with a uniform weight. If we examine the central interval, we perceive that the weight upon it is sustained by the resistance of the sections at m, p, and n, and the whole weight would be equal to the sum of the weights that each section separately would be capable of sustaining.

The resistance of each section being $R\, d^2$. If w represent the uniform weight upon the whole beam, we will have for the weight that the section m alone could sustain

$$R\, d^2 = \tfrac{1}{2} w \times \tfrac{1}{2} l = \text{ or } \quad w = \quad 4\,\frac{R\, d^2}{l}$$

For the section at n $\qquad w = \quad 4\,\dfrac{R\, d^2}{l}$

For the middle section $\qquad w = \quad 8\,\dfrac{R\, d^2}{l}$

And for the whole weight $\qquad\qquad 16\,\dfrac{R\, d^2}{l}$

which is twice the weight that the middle section alone is capable of sustaining, or in other words: The strength of a beam fixed at the ends is to the strength of a beam free at the ends as 2 is to 1.

For the end section ($n\ o$) we have weight which the section at n alone would sustain $\qquad w = 4\,\dfrac{R\,d^2}{l}$

Weight which the section at o would sustain $\quad w = 0$

Weight which the section at p' would sustain $\quad w = 8\,\dfrac{R\,d^2}{l}$

$$\text{Total} \quad 12\,\dfrac{R\,d^2}{l}$$

The strength of the end span, or of a beam fixed at one end and free at the other, is to the strength of a beam free at both ends as 12 : 8, or as 3 : 2.

When the span becomes considerable, simple timbers are insufficient, and framed trusses become necessary. Whatever may be their particular form, the object in every case obviously is, to dispose of a given quantity of material so as to resist effectually all the forces which tend to produce rupture or change of form.

The consideration of the case of a single beam involves the principles of a framed truss, the same forces act in both, the manner of resisting them alone is different: in the former, the cohesion of the fibres secures the object; in the latter, it must be attained by a judicious combination of ties and braces.

It has been shown, that in a beam the parts near the axis are but little strained, and consequently oppose but little resistance; hence, they could be removed without serious injury; and, if the same amount of material could be disposed at a greater distance from the axis, the strength and stiffness would be increased in exact proportion to the distance at which they could be made to act: hence, the first object in designing a truss, must be, to place the material to resist the horizontal forces at the greatest distance from the neutral axis, which the nature of the structure will allow.

It is evident, however, that if two longitudinal timbers should be placed parallel to each other, without intermediate connections, nothing would be gained; for, in this case, each would act independently of the other, and the strength would be less than that of a single beam. Neither would a connection by means of vertical ties, as in the figure, add to the strength; for, the weight of the ties would increase the load, to resist

Fig. 44.

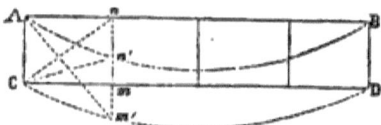

which, there would be only the stiffness and strength of the beams AB and CD.

By observing the effect of flexure upon this system, we are at once enabled to perceive the means by which it can be prevented.

The rectangles formed by the horizontal and vertical pieces are converted into oblique angled parallelograms, one diagonal of the rectangle, as Am, being lengthened, and the other, as Cn, shortened; and, as this effect must take place to a greater or less extent whenever any degree of flexure is produced, it may be concluded, that the introduction of braces which would prevent any change of figure in the rectangles will effectually prevent flexure. This is in fact the case, and the combination of timbers represented in figure 45 is sufficient to form a complete truss, capable, when properly proportioned, of resisting the action of any uniform load.

Fig. 45.

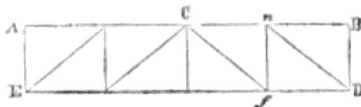

It appears, therefore, that in the construction of the vertical frame or truss of a bridge, at least, three series of timbers enter as indispensable elements; these may be called, chords, ties, and braces, and these are all that any uniform load requires.

The manner in which such a combination of parts acts to sustain a weight will now be examined.

CASE 1. Let the weight be uniformly distributed upon AB. It is evident that in case of flexure the depression will be greatest in the middle.

All the diagonals of the rectangles, in the direction of the braces, will have a tendency to shorten ; and, as this is effectually resisted by the braces, it follows, that such a truss is fully capable of sustaining a weight thus distributed.

CASE 2. Let the weight, instead of being uniformly distributed along $B E$, be applied at some point C'.

FIG. 46.

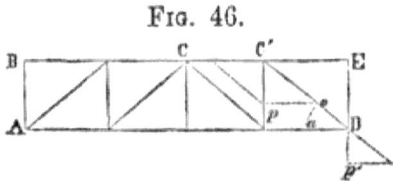

If we represent it by the portion $C' p$ of its line of direction, and construct the parallelogram of forces on $C' C$ and $C' D$, we find $C' o = w$ cosec. $a =$ strain on $C' D$.

If the point of application be removed to D, and again resolved into vertical and horizontal components, the vertical force will be equal to $D p' = w$. But this result is evidently false, for the weight is sustained by the points A and D, and presses upon them in proportion to the distances $C' B$ and $C' E$, it cannot therefore be equal to w at either. As cases of this kind frequently occur in attempting to trace the effects of forces upon the parts of a connected system, and often lead to error, we will endeavor to explain the cause of this apparently paradoxical result, which seems to contradict established principles.

If we suppose two inflexible rods, one horizontal and the other vertical, to be loaded with a weight applied at the angular point, A and D both resting against fixed points, then, the weight being represented by the portion $C p$ of its line of direction, may be resolved into components in the direction of $C A$ and $C D$.

FIG. 47.

The horizontal component can produce no vertical action at the point A, but, the oblique component, being transferred to the fixed point D and resolved into horizontal and vertical forces, will give a vertical pressure equal to Cp. This result is correct, because A sustains no vertical pressure and consequently bears no portion of the weight, which must, therefore, be thrown entirely upon D.

This case, however, does not present the true condition of the problem, for the upper chord does not abut against a fixed point at its extremity; if it did, the result would be correct, for it is evident, that a sufficient force applied at B, in the direction $B\,C$, would raise the truss, causing it to revolve around the point D, and relieving A of any portion of the pressure.

FIG. 48.

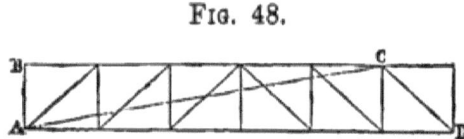

In consequence of the connection of the parts of the frame, the true line of direction will be $C\,A$, and the strains must be estimated by resolving the weight into components in the directions $C\,A$ and $C\,D$.

FIG. 49.

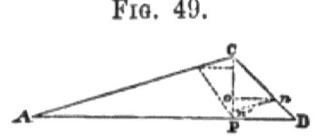

To simplify the demonstration, let the lines $A\,C$ and $C\,D$ be loaded at C, and the weight represented by Cn' resolved into components; Cn will represent the strain on CD, and $n\,n' =$ the strain on $A\,C$. By transferring the force Cn to the point D, and resolving it into horizontal and vertical components, we find the pressure on D to be equal to Co; and, in like manner, the pressure on A will be found to be $o\,n'$. But, from similar triangles, we have

WOODEN BRIDGES. 81

$$Co : Cp :: on : pD$$
and $$Cp : on' :: Ap : on$$

Hence, $Co : on' :: Ap : pD$ a result, which, as it makes the pressure upon A and D proportional to their distances from the line of application of the weight, must be correct.

We have seen, that in estimating the effect of a weight at C', it is necessary to resolve it into components in the directions $A\ C'$ and $B\ C'$.

In the same manner, it can be shown that the forces which act at C must by the connection of the system be transferred to the points B and A, in the directions of the diagonals $A\ C$ and $B\ C$.

FIG. 50.

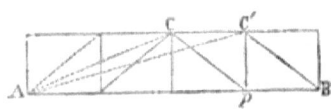

The effect of the oblique force $C'\ A$ upon the angle C evidently is to force it upwards, and the strain would be the diagonal of a parallelogram constructed upon $A\ C$ and $C\ C'$.

This result is of the greatest practical importance, and the existence of a force acting upwards appears to have been overlooked by many practical builders, as in some very important structures no means have been used to guard against its effects.

The consequence is, that in a straight as well as in an arched truss, a weight at **one** *side produces a tendency to rise at the other side.*

FIG. 51.

The effect of this upward force is to compress the diagonals in the direction of the dotted lines and extend them in the direction of the braces; but as the braces, from the manner in which they are usually connected with the frame, are not capa-

6

ble of opposing any force of extension, it follows, that the only resistance is that which is due to the weight and inertia of a part of the structure. When the load is uniform this is sufficient, because the weight on one side is balanced by an equal weight upon the other, and every part is in equilibrium. But, when the bridge is subjected to the action of a heavy weight, as a locomotive engine or a loaded car rapidly passing over it, and acting with impulsive energy upon every part at different instants, it is obvious, that no adequate resistance is offered by a truss composed of only the three series of timbers already described. Yet, we find that such a truss has been used for a large proportion of the bridges that have been erected, sometimes with, and sometimes without the addition of an arch, an appendage which, although it adds to the vertical strength, diminishes but little the effect of the force under consideration. No one, who has had an opportunity of observing it, can have failed to notice the great vibration produced in such bridges by the passage of a loaded vehicle. In long bridges, the undulations produced by the passage of a car can be felt at the distance of several spans.

The remedy for this defect is obvious; it is only necessary to prevent the diagonal, in the direction of the dotted line, from shortening, or, in the direction of the brace, from lengthening, and the upward force will be effectually resisted.

This requires, either, that counter-braces should be introduced in the directions of the dotted diagonals of the last figure, or, that the braces themselves should be capable of acting as ties, or, additional ties placed in the direction of the braces.

It follows, from the preceding exhibition of the effect of a variable load, that no bridge, either straight or arched, which is designed for the passage of vehicles, and particularly of railroad trains, should be constructed without counter-bracing or diagonal ties. It is only in aqueducts, where the load is always uniform, that they can with any propriety be omitted.

Effects of counter-bracing.

The consideration of the action of counter-braces leads to some very singular but important results.

Let the **truss** be loaded with a **weight so as to produce
some deflection, it has been** shown that **the diagonals in the
direction of the braces** will be **compressed, and in the direction
of the counter-braces** extended. **Suppose that the extension of
the** last named **diagonals is sufficient to leave an appreciable
interval between the end of the** counter-brace **and the joint
against** which it abuts, **and that** into this interval **a key, or
wedge of hard wood or iron, is tightly introduced: it is evident,
that upon the removal of the weight, the truss, by virtue of its
elasticity, would** tend **to regain its** original position; **but this it
cannot do, in consequence of the** wedges at **the ends of the
counter-braces, which prevent the** dotted **diagonals from recovering their original length, and the** truss is **therefore forcibly
held** in the position in which the weight left it; **the reaction of
the** counter-braces producing **the same effect that was** produced
by the weight, **and** continuing **the same strain upon the ties**
and braces.

The singular consequence necessarily results **from this, that
the passage of** a load produces **no** additional **strain upon any
of the timbers, but** actually leaves **some of them without any**
strain at **all.**

FIG. 52.

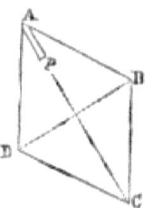

To render **the truth of this assertion more clear, we will**
confine ourselves to the consideration of **a single rectangle,** and
suppose **that** the effect of the flexure **caused by an** applied
weight has been to extend the **diagonal** $A\,C$ by a length equal
to $A\,p$, and to compress the brace $B\,D$ by an equal amount.

The point p **will evidently be drawn** away from A, leaving
the interval $A\,p$. **If a wedge be** tightly fitted into this interval
without being forcibly **driven,** it evidently can have no action

upon the frame so long as the weight continues; but upon the removal of the weight it becomes forcibly compressed, in consequence of the effort of the truss, by virtue of its elasticity, to return to its former position. This effort is resisted by the reaction of the wedge, which causes a strain upon the counter-brace $A\,C$ sufficient to counteract the elasticity of the truss; and, as no **change** of figure can take place, it follows, that the brace $B\,D$ cannot recover its original length, and, therefore, continues as much compressed as it was by the action of the weight.

The effect of a weight equal to that first applied will be to relieve the counter-brace $A\,C$, without adding in the slightest degree to the strain upon $B\,D$.

As regards the effects upon the chords, it is evident that the **strains are** only partial, and tend to counteract each other. **The** maximum strain in the centre is estimated by the force **which** would be required to hold the half truss in equilibrium **if the** other half be **removed; and** this **is dependent only on the** weight and dimensions of the truss. In **fact, if we** examine the parallelogram $A\,B\,C\,D$, we find that the effect **of** wedging the diagonals will **be to** produce strains acting **in** opposite directions at A **and** B, and destroying each other's effects; **the strains** produced **by** wedging any rectangle **cannot** therefore **be continued to** the next, and **of** course can have no **influence upon the maximum forces at the centre.**

As the **vibration** of **a** bridge is caused principally by the **effort to recover its** original figure after the compression produced **by a passing load,** it follows, that if this effort is resisted, **the vibration must be greatly** diminished, **or almost** entirely **destroyed.**

This **accounts for the surprising** stiffness which is found **to** result from a **well-arranged system of** counter-braces.

Inclination of braces.

From the preceding examination into the distribution of the **forces, we** learn that at least four sets of timbers are necessary **in every** complete and well-arranged truss.

The proper disposition, and the relative proportion of the **parts,** next demand attention.

The horizontal strain in the centre of the span, being **equal** to the force which would sustain the half truss in equilibrium, is independent **of the particular number or inclination of** the braces. **The same may be said of the pressure upon the abut**ments, which **is always proportional to the distance of the cen**tre of gravity from the point of support at the other **end of the** truss.

The parts of a frame can only act by distributing **the forces which are applied to** it,—they **cannot create force ; hence what**ever **be** the inclination of the braces, the pressure upon **the** abutment and the strain upon **the** centre of the chord must remain the same, with the same weight. It might be inferred, therefore, that the degree of inclination was of little consequence, or that different angles would be equally advantageous. That such is not the case can be rendered evident by the following considerations.

1. The braces must not be so long as to yield by lateral flexure.

2. The chords being unsupported in **the intervals between** the ties, these intervals must be limited by the condition that no injurious flexure shall be produced by the passage of a load.

On the other hand, as the ties approach each other, **the** angle of the brace increases; and when the intervals become small, the number of ties and braces is greatly increased, **and** with them the weight of the structure.

The true limit of the intervals can be readily determined when the size of the chords and the maximum load are known ; for it should evidently be such that when the load is at the middle, the flexure should not exceed a given amount.

The portion of the chord between any two ties is in the condition of a beam supported at the ends and loaded in **the** middle.

Should the angle of the brace as determined by this condition be too great, the remedy consists in introducing intermediate timbers as represented by the dotted lines in the marginal figure, and it is evident that by the addition of these we are enabled to vary the inclination at pleasure. A system of framing that will admit of the introduction of such timbers may therefore prove very advantageous,* provided they are not introduced at the sacrifice of counter-braces or diagonal ties.

FIG. 53.

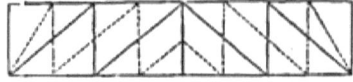

To determine the strain upon the counter-braces.

It is evident in the first place that counter-braces do not assist in sustaining the weight of the structure; on the contrary, the greater the weight and the degree of flexure which it occasions, the more will the counter-braces be relieved.

The strain under consideration must therefore result from the variable load, and as the effect of a weight on one side of a truss is to produce a tendency to rise at the opposite side, which is resisted by the counter-braces; and as this strain is not confined to a single timber, but distributed amongst several; as it is also resisted by the weight of portions of the structure, modified by the nature of the connections and the degree of flexure which the timbers experience, it would be a very complicated problem to trace the effects of a weight through the system of timbers which compose the truss, in order to determine the maximum strain upon a single timber. Fortunately we have a more direct and accurate means of obtaining the result.

It has been shown that by driving a wedge at the joint of the counter-brace, a permanent strain may be thrown upon the brace equal to that which results from the passage of the maximum load; this strain is not increased by the passage of the load, the effect of which is simply to relieve the counter-brace, and as the compression of the brace if within the elastic limits is in no respect injurious, but on the contrary highly conducive to stiffness, it follows that the compression of the counter-brace to an extent sufficient to throw a strain upon the brace equal to that which results from the passage of the maximum load,

---

* See description of the improved lattice.

is not only admissible, but very desirable. This strain is the greatest that can be thrown upon a counter-brace; the passage of a load relieves instead of increasing it, and it will be safe to calculate the size by the condition that it shall produce the required compression on the brace.

Let $ABCD$ be a rectangle, and suppose a force to act in the direction of the diagonal AC, C being a fixed point.

Fig. 54.

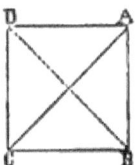

If the intensity of the force be represented by AC, the components in the direction of the sides will be AD and AB, and those which result from the resistance of the fixed point C will be CD and CB. These four components produce a force of extension on the diagonal DB, the magnitude of which is represented by DB. This is the measure of the force which must be produced by wedging the counter-brace; and as this diagonal is equal to AC, it follows, that the strain upon the counter-brace which produces a given pressure upon the brace, is equal to that pressure itself.

If w represent the greatest weight that can ever press upon any point of a bridge, $W \times$ Sec. BAC will be a little greater than the strain, and in practice may be taken to represent it.

As the greatest accidental weight that can ever act at a single point is very small when compared with the uniform load, it follows, that counter-braces may be very small when compared with other timbers.

To determine the strain upon the braces and ties.

To estimate the strain upon the brace Df, we may suppose the whole of the bridge between A and D to be suspended at the point D, and the measure of the force would be that which would hold it in equilibrium; but in estimating the weight, it is not sufficient to take simply the weight of the structure itself, to this must be added the greatest load that could ever come upon it.

Fig. 55.

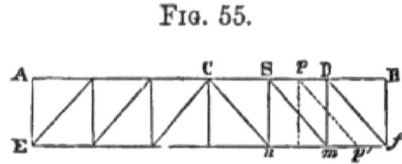

In a railroad bridge, the greatest load is probably when a train of loaded cars occupies the entire space between A and D, and the driving wheels of the locomotive are directly over D. The weight of the cars may be regarded as a uniform load distributed over $A\,D$, and its centre of gravity would be at a distance of one-half $A\,D$ from the point of rotation E. The weight on the driving wheels of the locomotive may be considered as acting with a leverage equal to the whole distance $A\,D$. Let the weight of the bridge between A and D be represented by W, the uniform load on $A\,D$ by w', and the weight on D by w''; then taking the sum of the moments, we have $w\,(\tfrac{1}{2}\,A\,D) + w'\,(\tfrac{1}{2}\,A\,D) + w''\,(A\,D) = f\,A\,D$ or $\dfrac{w + w' + 2\,w''}{2} = f =$ the vertical force which applied at D will sustain the load.

The strain upon the brace will be very nearly $f\;sec.\;m\,D$ f or $f \times \dfrac{D\,f}{D\,m}$.

When there are intermediate braces and ties, as $p\,p'$, it will not vary much from the truth to suppose the strain which was thrown upon a single brace in former case by the uniform load to be divided equally amongst all that the interval contains. If one intermediate tie be introduced, it will bear one-half, if two, each one-third, if n, $\dfrac{1}{n+1}$ part of the uniform load, and this is expressed by $\dfrac{w + w'}{2\,(n+1)}$.

The weight on the driving wheels of the locomotive **being applied** at a single point, could **not be regarded as divided** amongst all the intermediate braces of the interval $S\ D$. When this weight is at p, the brace $p\ p'$ will sustain more of the pressure than $S\ m$, or $D\ f$. The proportion will depend on the **stiffness of** the chord and the compressibility of the **brace, and** must be determined upon the principles by which **we ascertain** the pressure of a beam laid over several supports.

The problem for determining **the** strain thrown upon a particular brace by the passage of a variable **load,** is very complicated in its practical application, and in consequence of defects of workmanship, little confidence can be placed in the results.

Even if we suppose the whole weight w to be thrown upon the intermediate **brace without** the adjacent ones sustaining **any** portion of it, the area will not be increased more than from three to five square inches above the **true dimensions ; and in** practice this allowance might be made without increasing **much** the size of the timbers, although it is **evidently more than** sufficient.

Of the strain upon the ties and braces at the centre.

We have seen, in examining the forces **which** act upon a single beam supported at the ends and uniformly loaded, that there exist both horizontal and vertical forces at every point except **at the middle and at the** extremities. At the middle, when the load is uniform, the strains are altogether horizontal, **the** two parts being exactly balanced, mutually support each **other,** and no vertical **strain** is experienced ; but at other points, **the** distances to the extremities being unequal, the pressure of one part will be greater than that of the other, the horizontal strains will no longer balance, and the difference must be compensated by the vertical resistance produced by a cross strain upon the **fibres.**

This vertical force increases from **the** centre, where it is zero, to the extremities, where it is equal to one-half the whole uniform weight upon the bridge ; and the increase, when the weight is uniform, is proportional to the distance from the centre

In a bridge, the office of the chords is to resist the horizontal forces, and that of the ties and braces the vertical forces, and as the strain resulting from the uniform load is zero at the centre, it follows, that the sizes of the intermediate timbers may be much smaller here than at the abutments, as they have very little more strain to bear than that which results from the portion of the variable load, which acts immediately over them, which, in a long span, is comparatively trifling.

Each successive brace, in passing from the centre to the abutment, is more and more strained, and consequently should, if properly proportioned, be increased in size, but as such increase would add greatly to the expense and trouble of framing, it is preferable in practice to make all the timbers uniform and compensate for the additional strain at the ends by additional braces called arch braces.

As the preceding method of investigation might be considered objectionable, and doubts be entertained of the correctness of the important consequences which result from it, on the ground that the analogy is not perfect between a beam supported at the ends and the framed truss of a bridge, we will endeavor to present a different view of the subject.

The important principle that we aim to establish is, that a great difference exists between the strains on the ties and braces at the centre and at the ends, the precise law of increase or diminution is of secondary importance, and will not now be considered.*

It has been stated that when a truss settles, the rectangles formed by the ties and chords become oblique parallelograms, the diagonal in the direction of the brace being compressed and the opposite one extended. Could we ascertain the exact degree of reduction which the length of the brace experienced, we would have a certain measure of the strain. To determine this by calculation would be difficult, as the change of figure

---

* This paragraph was penned about eight years ago, at which time the writer was not aware that any bridges had been constructed in such a way as to recognize the existence of this increased strain at the abutment. But in several of the plans that are now in use the principle seems to have received attention.

would be affected by the form of the curve of flexure, and the changes in the lengths of the sides of the rectangles.

FIG. 56.

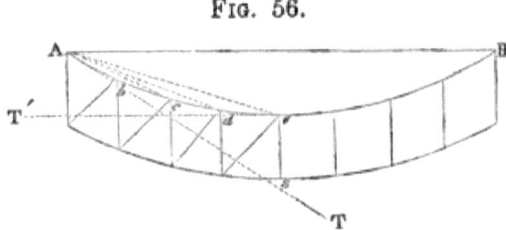

We can however approximate to the result sufficiently near to confirm the previous conclusions.

Regarding the curve of flexure when very slight as a circle, draw lines to all the angular points, and AT a tangent at the point A. Call $a =$ angle TAb and $n =$ the whole number of intervals Ab, bc, cd. As the chords are all equal, the angles at A would also be equal, and the angle $TAB = na$, which measures the angle of change of the first rectangle, is n times as great as $T'ed = a$, which measures the angle of change of the rectangles in the centre. Now, for small deflections, the diminution in length will be proportional to the angle, and, as this diminution measures the strain, it follows, that it must be n times as great at the abutments as at the centre. When n becomes infinite, as we may consider it in a beam, $a = o$, and the strain at the centre is nothing.

The changes in length of the chords in the central interval do not affect the diagonal, as they compensate each other.

The strain at the abutment is constant whatever may be the length of the intervals, as it is measured by the tangent BAT: results which correspond with those obtained by the other method.

Effects upon the braces and ties which result from the introduction of Arch-braces.

In a system of this kind it is necessary to estimate the strains by dividing the truss into parts, and considering the action of each part separately.

Let dD and eE represent **two arch-braces** extending from the points d and c near the centre **to the** abutments. If we disregard compression, the effects of which will be considered subsequently, it is evident that the weight of the portion dc may be regarded as suspended from the points d and c, and will be entirely transmitted to the abutments, **exerting no influence whatever** upon the parts Ad and cB.

FIG. 57.

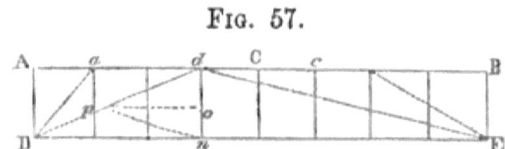

If aD be another arch-brace, the portion ad will be suspended from a and **d,** and its weight transmitted to the abutments by aD and dD.

The calculation of the strains **in this case, therefore,** becomes extremely simple; we can regard **the whole weight of the bridge as supported by the arch-braces, and the load upon** the ordinary braces will be only that which **is due to the small** intervals ad, dc; at d for example the **strain upon the ordinary** brace would be one-half the weight on dc, at a **it would** be one-half the weight on ad, and it therefore follows, that by the introduction of arch-braces of sufficient size to make their compression inconsiderable, the ordinary timbers may be **reduced to very small dimensions.**

The weight upon any arch-brace (dD) is one-half the load upon $(ad + dc)$. Call this weight w, and let $DE = s$, $dn = h$, $Dn = m$, $nE = m'$, and $l =$ length of brace $Dd;$

Then $s : m' :: w : \dfrac{w m'}{s} = do =$ portion of the weight transmitted to D.

Also $h : m :: \dfrac{w m'}{s} : dp = w \dfrac{m m'}{s h} =$ strain upon the brace bD. This strain is a maximum when $m = m'$ or when b is at the centre.

We cannot however regard the arch-brace as incompressible; **on the contrary, it is** known that timber will admit of a reduction of .001 of its length without impairing its **elasticity, and therefore an arch-brace 100 feet** long would admit of a reduction **of 1⅛** inches **without** injury.

The **effect of this compression would be to throw a portion** of the strain upon the ordinary ties and braces depending **upon** the relative **stiffness of** the truss, **and the compressibility of the** arch-brace.

As the truss is relieved in proportion to the amount of strain thrown upon the arch-brace, the introduction of wedges at the extremity would furnish the means of regulating this proportion **at** pleasure, **and,** if necessary, the arch-braces could be **made** to sustain the whole of the weight.

It may not **be** improper at this place to advert to an objection which is sometimes made to a combination of two distinct systems in **the** same truss. **It is said that** the workmanship can never be so perfect that **both systems will assist** in sustaining the load, and **that either one** or **the other will** sustain the whole. **This** remark was made by a gentleman **who** deservedly **ranks** amongst the first in his profession **as a Civil Engineer, and who is the** inventor of a plan of bridge construction which **the writer regards as one of the most scientifically** proportioned **of any in general use.** That it is erroneous we think **can** be readily shown.

Were the materials of a bridge perfectly incompressible, **or even nearly so,** the remark would be correct, **as** one system **would break before** the other could be **brought into** action. But **wood** is highly elastic, and admits of a considerable extension **or** reduction of length without injury; consequently, if at the instant of striking the false works, one system should be overloaded, it would settle more than the other, and the second be thus brought into action. The strain upon the two might **not be** precisely **equal; but this** is of little practical **consequence.**

Experience confirms this view: many of the bridges on the public works of Pennsylvania, having been too lightly proportioned, settled greatly; they were strengthened by the addition of arch-braces or arches, and have since stood well. If then the introduction **of an** independent system, after a truss has commenced to yield, can arrest its progress, it cannot be doubted that the effect would be still more beneficial if introduced at the time of its construction.

To determine the strains upon the chords.

The maximum strain upon the chords of a straight bridge is in the centre; being one of compression on the upper chord, and of extension on the lower, its magnitude is represented by a force which, applied horizontally at A, would keep the half truss $A\,B$ from falling.

If the bridge be uniformly loaded, the centre of gravity will be vertically under n, the middle point of $A\,B$, and if w represent the uniform weight, its moment in reference to the point of rotation C will be $w \times m\,c$, or if s represent the span it will be $\dfrac{w\,s}{4}$. An accidental load will produce the greatest strain at the centre; its leverage will then be $\tfrac{1}{2} s$ and $w' \times \tfrac{1}{2} s = \dfrac{w'\,s}{2}$ = moment of the accidental load.

FIG. 58.

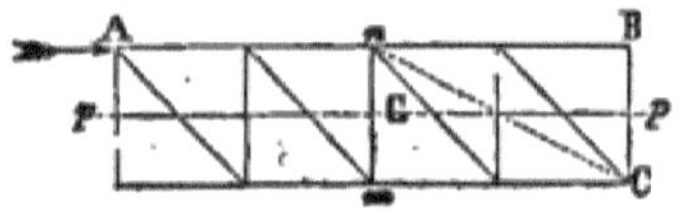

The sum of these moments will be $\dfrac{w + 2\,w'}{4}\,s$.

The horizontal force at A acts with a leverage $B\,C = h$, its moment will therefore be $H\,h$. Hence $H = \dfrac{w + 2\,w'}{4\,h}\,s =$ the force which measures the tension at the lower, and the compression at the upper chord.

The same result would be obtained by referring the moments to a neutral axis passing through the middle of the truss; and although not quite so simple, this is in some respects the best way of considering the question, as there are in fact two forces, one at the upper, the other at the lower chord, acting with a leverage $= \frac{1}{2} h$. The analogy of the truss to a beam supported at the ends is thus preserved.

FIG. 59.

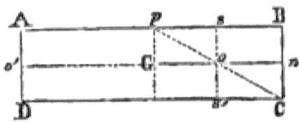

If we suppose a single force acting at A to keep the truss in equilibrium, C being a fixed point and (G) the centre of gravity, we may transfer the force at A to the point p of its line of direction, and the two forces $p B$ and $p G$ will have a resultant, which, in the case of equilibrium, will pass through the point C, and $p C$ will represent the direction of the force.

We can, however, estimate the moments in a different manner, and one which, although it will give the same result, will express better the true conditions of the problem.

Instead of one force there are two, one at A, the other at D. The effect of these forces would be to cause rotation around the middle of the line joining their points of application. The locus of the point of rotation must therefore be in the line $o' n$. But the weight of the truss reacts upon the fixed point C, and generates a resistance which can be replaced by a force acting upwards. There are, therefore, two vertical as well as two horizontal forces, one acting upwards at C, the other downwards through the centre of gravity G. As these forces are equal, the locus will be in the line $s s$ which bisects $G n$, and the intersection of the two loci $s s'$ and $o' n$ will give o, the true point of rotation of the four forces. The resultant in the case of equilibrium passes through o and C, and evidently coincides with the line $c p$ as first determined.

FIG. 60.

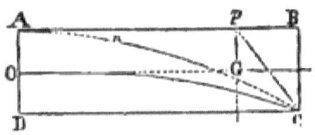

If we suppose the truss to be without a horizontal tie, the conditions of equilibrium are in no manner affected: the resistance of the abutment at C supplies the place of the tension at D, and the direction of the resultant Cp, which represents the pressure upon the abutment, will be determined as before.

Should it be desirable to construct a curved truss on the principles of the arch of equilibrium, instead of making $O\,C$ the proper figure for this curve, it would be better to bolt it to the truss in the direction of the line $A\,n\,C$. $O\,C$ could then be made of any form that would produce the best architectural effect, and the curve of equilibrium, in consequence of the greater rise, would be much increased in strength. The curve of equilibrium is useful only where the road is constant, or the variable load relatively very small.

The weight of a structure can be ascertained from the bill of materials; and the readiest way of determining the position of the centre of gravity in any intricate combination is by means of a model. It is not however necessary in ordinary cases to resort to this. The weight of the roadway, as also the maximum load upon it, is nearly uniformly distributed, so also are the weights of the chords; the only increase in weight towards the abutments is due to the increased lengths of the ties and braces, which being very small when compared with the other weights, can affect but slightly the position of the centre of gravity; and as the small error is in favor of stability, it is in almost all cases proper to estimate the centre of gravity at a distance from the abutment equal to one-fourth the span.

Where the chords are of considerable depths, it becomes necessary to take into consideration the distance from the neutral axis.

FIG. 61.

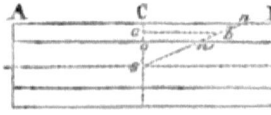

As the strains vary as the distance from the neutral axis, if we represent the strain at C by Cn, and join $n\,s$, $o\,n'$ will represent the strain at o and the whole resistance of the chord will be represented by the area of the trapezoid $C n\, n'\, o$ multiplied by the distance of the centre of gravity from s.

The centre of gravity will always fall nearly in the middle of $C o$. Even when $C o = o\, s$, the error will be only $\frac{1}{13}$ of $C o$, by considering it in the centre, and in ordinary cases it will not be $\frac{1}{80}$. The error is, moreover, in favor of stability.

We may therefore, in practice, consider the centre of gravity as falling in the middle of $C o$, and the leverage will be the distance from this point to s. The average strain upon the joint is represented by $a\, b$, and is determined from the proportion $s\, c : s\, a :: R : a\, b = R\, \dfrac{s\, a}{s\, c}$. R representing the maximum strain that it is considered safe to allow.

Sometimes two or more chords are used at different distances from the neutral axis.

FIG. 62.

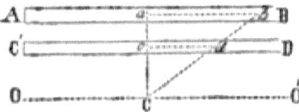

Let $A B$ and $C D$ represent two chords, at distances $a\, C$ and $c\, C$ from the neutral axis $o\, o'$. Draw $a\, b$, to represent the maximum strain proper to allow at the centre of the upper chord, and join $b\, c$, $c\, d$ will then represent the strain upon the second chord, which accordingly opposes much less resistance than the first.

As this case often occurs, particularly in the construction of ordinary lattice bridges, it may be satisfactory to give the equation of moments. Let $R =$ strain per square inch, at the distance $a\, c$, $a =$ area in inches of the section of each of the four chords, $d = a\, c$, $c = c\, C$. Then $R\, \dfrac{c}{d} = C\, d =$ strain per square inch on second chord. $a\, R \times d =$ moment of first chord. $a\, R\, \dfrac{c}{d} \times c =$ moment of second chord.

$a R (d + \dfrac{c^2}{d}) = \dfrac{w s}{8}$ is the equation of moments, from which the strain upon the chords with a given weight can be calculated, or, the strain being assumed, the size of the cross-section (a) can be ascertained. In this expression w = whole uniform weight, s = span.

If a truss be constructed with parallel arches, the strains upon each will be calculated upon the same principles.

FIG. 63.

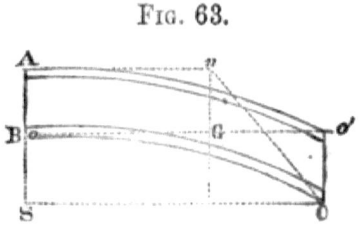

Let $A o'$ and $B C$ be two curved chords or arches. The resistance of the abutment at C, takes the place of the strain upon the lower chord, in a straight bridge, and the neutral axis will still exist half way between A and S. The horizontal strains varying as the distance from the neutral axis, it follows, that the two arches can never sustain equal portions of the strain; but if one of them, as in the figure, intersects the line $A S$ at its middle point, it will not assist in the slightest degree in sustaining the strain at that part of the truss, which will be thrown entirely on the upper arch at the point A.

At the abutments the condition of things is reversed; the lower chord sustains a horizontal strain equal to that at the centre, and, in addition, a vertical force resulting from the weight of the half truss. The resultant of the two is an oblique line $c n$, determined by joining C with the point of intersection of a horizontal line through A, and a vertical through the centre of gravity G. The end o' of the upper chord sustains no pressure.

This case is one in which apparent strength is real weakness; one of the arches at the centre, and the other at the abutments, contribute nothing towards sustaining the horizontal strains. A single arch extending from A to C, would give the same strength with half the material, when the dimensions are uniform throughout.

That a system similar to that represented in the figure should be properly proportioned, the upper arch should diminish from the centre to the end, and the lower one from the end to the centre.

We will conclude this part of the subject with a practical exemplification of the manner in which the sizes of the chords are calculated.

As our object is merely to illustrate a principle, great accuracy will not be attempted, and round numbers only will be used.

FIG. 64.

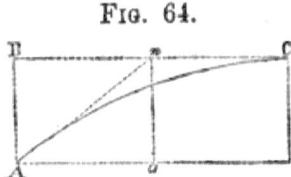

Let x = area of the cross section of the arch in inches
s = span = 500 feet
w = whole weight
$\frac{s}{10}$ = rise of arch = 50 feet
R = 1000 lbs. = maximum strain per square inch.

$$R\, x\left(\frac{s}{10}\right) = \frac{w}{2} \times \frac{s}{4} \text{ whence } x = \frac{10\,w}{8\,R} = \frac{5}{4}\cdot\frac{w}{R}.$$

The greatest variable load is generally considered as caused by a crowd of people, and is estimated at 120 lbs. per square foot.

If the bridge be supposed 20 feet wide, and supported by four trusses, each will bear $5 \times 120 = 600$ lbs. per foot linear. The weight of the structure must be determined by assuming the dimensions of all its parts, making out a bill of material, and finding its weight from a knowledge of its cubic content and specific gravity. At present, we will assume the weight of the structure to be equal to the greatest load, 600 lbs. per foot.

We then have $W = 1200 \times 500 = 600,000$ lbs.

The horizontal strain at C is $R\,x = \dfrac{10\,w}{8} = 750{,}000$ and $a = 750$ square inches, or about 27 inches square for the size of the arch at the centre.

We have also $\dfrac{w}{2} \times \dfrac{A\,n}{n\,o} =$ strain at $A = \sqrt{\dfrac{\frac{s^2}{100} + \frac{s^2}{16}}{\frac{s}{10}}} \times$

$\dfrac{w}{2} = \dfrac{w}{8}\sqrt{116} = 802{,}500$ lbs. or 802 square inches, about $29\frac{3}{4}$ inches square for the size of the arch at the abutments.

When two independent systems are combined in the construction of a truss, it becomes difficult, if not practically impossible, to estimate the portion of the strains sustained by each, owing to the defects in mechanical execution inseparably connected with every structure. If a calculation be attempted, it can only be upon the supposition that the joints are absolutely perfect, and that at the first instant of flexure both systems are in full bearing, and oppose a resistance proportionate to their relative stiffness.

Disregarding the particular arrangement of ties and braces, or the greater or smaller number of the intervals, we will consider the trusses as acting as a whole. This can be done with propriety on the supposition that the joints are perfect, and a general solution of the problem becomes very simple.

If the trusses act as a whole, the deflections may be considered as proportional to the weights; but the strains upon the chords are as the weights directly, and as the areas of the cross-sections inversely, and the deflections must therefore be in the same proportion.

Let a represent the area of the cross-section of the chords of one system, and $n\,a$ that of the other: the depth and length of truss in each being equal. If d represent the deflection of the first system, with a given weight, $n\,d$ will express the deflection of the second.

Let x represent the actual deflection, which is of course equal in both. Then $d : x :: w : \dfrac{w\,x}{d} =$ weight on first system, and $n\,d : x :: w : \dfrac{w\,x}{n\,d}$ = weight on second system. The sum of these must equal the whole weight. Hence $w = \dfrac{w\,x}{n\,d} + \dfrac{w\,x}{d}$. Whence $x = \dfrac{n\,d}{n+1}$ and $\dfrac{w\,x}{d} =$ $\left(\dfrac{n}{n+1}\right) w$ = weight on first system, $\dfrac{w\,x}{n\,d} = \dfrac{1}{n+1} w$ = weight on second system.

In other words, the strain upon each system will be exactly proportioned to its powers of resistance, and the whole together may be estimated as one truss.

In the construction of a bridge with a system of arch-braces, the simplest and best plan is to depend upon the latter to sustain the entire weight of the structure, using only a very light truss with counter-braces or diagonal ties to establish the necessary connection of the parts, prevent flexure and vibration, and resist the action of variable loads.

Instead of using arch braces, trusses are sometimes strengthened by the addition of arches. Great benefit results from their use, but nearly the same effects may be obtained by arch-braces.

An arch, when of the proper figure of the curve of equilibrium, is capable of sustaining any constant load without change of form; but, as the load upon a bridge is variable, it is obviously impossible to make an arch of equilibrium for a wooden viaduct.

The flexibility of an arch renders it but poorly adapted to sustain a variable load; when used for this purpose, therefore, it must always be connected with a truss capable of giving it the necessary stiffness. Such combinations are extensively used.

Means of increasing the strength of bridge trusses.

When a truss, in consequence of having been too lightly proportioned, gives way by vertical flexure, an arch, or arch-braces with a straining-beam connecting the upper ends, may be bolted to the truss. Such additions have been often made, and are found to answer well. Many of the bridges on the public works of Pennsylvania **have been strengthened in this** way, and rendered sufficiently **strong for the heaviest locomo-** tives and trains.

When a beam is laid **over** several supports, its strength **for a given interval** is much greater than when simply **supported at the ends.** **The same** principle is applicable to bridges, and **when several** spans occur in succession, it is of great advantage to continue the upper and lower chords, if the bridge is straight, across the piers. By this arrangement, the strength of chords of each central span in a series would be double that of the **same** spans disconnected, and the extreme spans would be stronger in the proportion of 3 to 2.

Notwithstanding this, we often see bridges in which the upper chords are not connected over the piers, and the absurd remark has been made, by practical builders, that the bridge must yield somewhere, and better there than elsewhere. Just in proportion as this point is capable of opposing a resistance, must the strength of the bridge be increased; and it **is obvious** that if a bridge should be cut in **two in** the centre of the span, and one-half removed, the other half could **not fall as long as the** connection over the pier remained perfect.

Even **in** bridges of a single span, it would not be impossible to communicate the strength of a continuous bridge, by connecting the upper chords with chains passing over the back of the abutments, and anchored into the ground on the principle **of a suspension** bridge; but such **an arrangement** is not to **be recommended in ordinary cases.**

When the chords of a straight bridge are of equal size, the lower are necessarily much weaker than the upper within the elastic limits; the latter resist a force of compression which naturally **closes the joints, and** brings **every** part of the cross-section into full bearing. **But** the case is very different **at the** lower chord; here, from the nature of the strain, which is one of extension, the joints are opened, and, from the manner in which the connection is formed, only one-half the area of the **cross-section** opposes any resistance.

Fortunately, we have a simple means of correcting the evil; **but, simple as it is,** it does not appear to be generally resorted to. It consists in placing the ends of the lower chords in close contact with the abutment, or, which is still better, driving wedges between the abutment and chords.

Fig. 65.

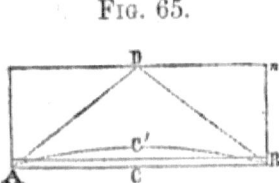

In fact, it is evident that if the truss rests against an abutment capable of opposing a horizontal resistance, the tie $A B$ could be cut entirely through without danger of the truss falling, for the strain upon the tie at C is exactly replaced by the two resistances to compression at A and B; and just in proportion as a pressure can be produced at A and B, in exactly the same proportion will the truss be relieved; and if this pressure should, by wedging or other means, be made to exceed the horizontal thrust of the truss, the centre D would be forced upwards.

If the chord $A B$ should be slightly curved, as represented by the dotted line $A C' B$, the result would be the same; the pressure upon the abutment would not become almost infinite, as was asserted by a gentleman with whom the author once corresponded upon the subject of bridge construction. Whatever might be the rise $C C'$, the strain at A and B would not be in the slightest degree affected so long as the weight, span, and height $C D$, or its equal $n B$, remained constant.

A moderate pressure upon the face of the abutment, so far from being an injury, is a very decided advantage, as it serves to counteract the pressure of the embankment on the other side, and allows a reduction in the thickness of the wall.

For the reasons assigned, we think that wedges behind the ends of the lower chords of straight bridges should never be omitted.

On the maximum span of a wooden bridge.

It requires no demonstration to prove that, in order that the maximum span may be attained in a bridge, it is necessary that every part should be properly proportioned to the strain that it may be required to bear. The strength of a system is the strength of its weakest point; this is the point of fracture; and any increase of strength at other points, produced by increasing the amount of material beyond its minimum quantity, only increases the weakness by increasing the weight. It follows, therefore, that no plan which does not distinctly recognise this principle of accurately proportioning the dimensions to the strains, and apply it in detail, can be employed for the maximum span.

Many large bridges have been constructed, several of which have considerably exceeded 300 feet in span; but in all these were some defects, some points too heavily loaded by timbers of unnecessarily large dimensions.

Tredgold, in his treatise on carpentry, gives a plan for a bridge of 400 feet span, the support of which consists of framed voussoirs, as they are termed; and as no mention is made of any variation in size, it was no doubt the design of the architect to make the dimensions uniform.

Fig. 66.

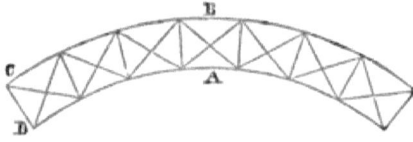

The defects of this arrangement naturally appear from the preceding explanation of the manner in which the pressures are distributed, varying as the distance from the neutral axis.

The points A and C sustain less than B and D, and if the sections are everywhere the same, it follows, that if B and D are sufficiently strong, A and C must possess surplus strength, and with it unnecessary weight of material.

In the second place, the diagonal timbers in the intervals add considerably to the weight, without corresponding advantage; an arch is very well able to resist a variable load upon one side, if the other side can be kept from rising; and instead of using a braced arch, it is better to make the truss with which it is connected serve as a counter-brace.

A system which appears to be absolutely the lightest and most simple that could be used, and the one of all others best calculated to attain the maximum limit of the span, is represented in the following figure.

Fig. 67.

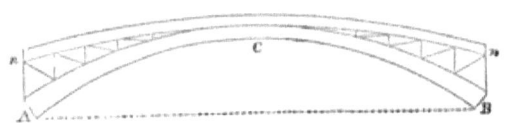

The sole support of this truss is the single arch AB, which increases in size from the centre to the ends in exact proportion to the strain; nn' represents the line of roadway, supported by uprights from the arch, and between these uprights keyed counter-braces or diagonal bolts are introduced. By wedging the counter-braces or screwing the nuts, the arch can be compressed sufficiently to produce a strain equal to that caused by the maximum load, so that the subsequent passage of a load will only relieve the counter-brace without adding to the weight upon the arch, and the upward motion at the haunches is effectually prevented by the counter-bracing.

The hand rail on the top of the bridge may, by a proper connection with the truss, be made to assist in counter-bracing the centre, and thus every part performs important functions without a single stick being superfluous.

An assertion incidentally made in the above description, perhaps requires further elucidation. We have said that, by wedging the counter-braces, a strain can be thrown upon the bridge equal to that produced by the maximum variable load, and that the subsequent passage of this load will throw no additional strain or weight upon the arch.

The importance of this fact renders it worthy of a **separate** consideration.

Effects of counter-bracing **upon an arch.**

Fig. 68.

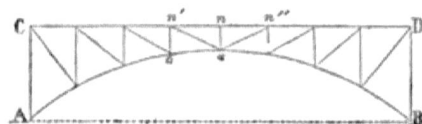

Let **A s B represent an** arch, supported by **resisting** abutments at the points **A** and *B*, and suppose a heavy uniform load to be distributed along the roadway *C D*. The effect of this load is to depress the arch, and the amount of the depression will be greatest in the centre, diminishing towards the abutments, where it is zero. In consequence of this difference of depression, any point (n) **near the centre will sink more** than (n') nearer the abutment; by this change of figure **the** diagonal n' *s* is lengthened, and n *o* is shortened. Consequently, if, before the application of the weight, the counter-brace n' *s* was exactly in contact with the joint, it must now be at such a distance as to leave an interval.

When the weight is removed, the arch will return to its original figure, and the interval will be closed; but if wedges be driven into all the intervals, their reaction when pressed **will** prevent the arch from regaining its figure, and it is forcibly held in the position in which the load placed it, and, as a necessary consequence, continues subject to the same strain as when the weight was upon it. It certainly needs no lengthy explanation to convince any one that, if a spring or a beam be bent, either by a weight or by the resistance of fixed points, if the flexure is constant the strains must be precisely the same, and, consequently, the counter-braced arch is not more strained when the weight is upon it, than it is when that weight is removed, the effect of the weight being simply to relieve the counter-braces, which are not strained when it is acting.

If any should regard the above explanation as unsatisfactory, and be unable to reconcile the apparently paradoxical result, that a heavy weight brought upon an arch produces no strain, the following considerations may serve to remove the difficulty.

We admit that the proposition is not strictly true, but it is nearly so, and would be entirely so, if the counter-braces and the longitudinal timber with which they are connected at top were incompressible; as they are not, the arch will rise a little after the removal of the weight, but, its elastic movements being confined within very narrow limits, great stiffness will be secured.

The upward action of the arch, in consequence of its elasticity, produces a force which presses against the counter-braces, and is by them transmitted to the longitudinal timber CD, upon which it produces a force of extension exactly the reverse effect to that which is produced by a weight; and when the weight acts, this strain is counteracted, and becomes nothing.

If it be asked, why does not the strain upon the arch throw an additional strain upon the abutments? The answer is, that this would be the case if the counter-braces could act against fixed points not connected with the frame itself; but, as this is not and cannot be the case, it follows, that the downward pressure upon the arch is exactly counterbalanced by an upward pressure at the other end of the counter-brace. When the weight is upon the bridge, the upward pressure is counteracted, whilst the downward pressure is as before, and therefore the pressure upon the abutment is increased by exactly this amount; that is, by the accidental weight, whatever it may be.

Instead of counter-braces of wood, diagonal ties of iron, with nuts and screws, could be placed in the direction of the other diagonals; and their use would be in some respects very advantageous. They would be lighter, and consequently would add less to the weight of the structure. The strain upon them being tensile, they could be of any desired length without danger of flexure, to which counter-braces are liable; and the degree of tension could be very conveniently regulated by means of the screws, without the danger of loosening, which is connected with the use of wedges.

Fig. 69.

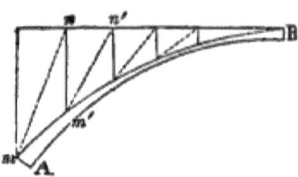

A truss of this kind with a parabolic arch, the section at every point being proportional to the strain, and protected from the effects of partial loads by the iron **diagonal** ties $m\ n\ m'\ n'$, &c., is absolutely the lightest that we can conceive for a wooden **bridge,** fulfils every condition of a perfect structure, and consequently **admits of** the greatest possible extension of the span. If a horizontal tie is desired, **the posts must be extended.**

Roadway.

The **roadway of a** bridge admits **of little** variation. It is generally constructed by laying beams across the trusses, **upon** which are placed the longitudinal pieces which carry the planking. A very important part of the roadway consists in the bracing, which is necessary to prevent lateral flexure. The usual arrangement of braces is shown in the annexed figure.

Fig. 70.

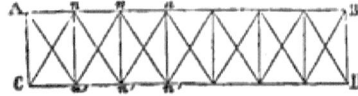

$A\ B$ and $C\ D$ are the trusses, $n\ n'$ the girders, and the diagonal timbers are braces.

The following figure represents another plan of horizontal

Fig. 71.

bracing, which is perhaps the lightest that could be used, and would be well adapted to large spans, where the quantity of material in the centre is required to be the least possible. The diagonal timbers between the arches are designed to be employed as keyed counter-braces.

A very good bracing may sometimes be obtained by spiking the floor plank in two layers, extending diagonally across the bridge and crossing each other at right angles.

CAST-IRON BRIDGES.

It is foreign to the original design of this treatise to introduce the subject of cast-iron structures; but as the same general principles must guide the engineer in these, as in other bridges, a paragraph upon the subject may not be considered out of place.

The abundance of wood, and its great relative economy, have secured its adoption in this country, in preference to iron; but in **Great** Britain, many splendid **structures** have been erected of the **latter** material, which possess great beauty, **strength,** and durability.

If the principle of proportioning every part to the strain which it has to bear, is important in its application to timber-bridges, much more must it be when applied to bridges of cast-iron; for the expense is nearly in proportion to the quantity of material, and **the weight, and** consequently the weakness, is increased by **every pound unnecessarily added.** As we have said already, **the strength of a bridge is** the strength of its weakest point; and **of course the** accumulation of **material where it is not needed,** so far **from being of advantage, is a positive injury.**

It is therefore of the first importance, in designing a plan for a cast-iron bridge, to place the material which is to resist the horizontal strains at the greatest possible distance from the neutral axis, as it will there act with the greatest effect. This object is secured by using only a single arch, and giving it the maximum rise that the nature of the structure will admit. The only objection that can possibly be made to a single arch, we conceive to be its flexibility; but if it can be so counter-braced as to prevent a change of figure by the action of a variable load, we cannot perceive that any thing more is necessary.

If this principle be correct, it follows, that most of the plans which have been used are to some extent objectionable, as they consist either of framed voussoirs, that is, of two parallel arches separated by cross-braces, or of several arches rising at different heights, and extending to different elevations. The latter arrangement would perhaps be a good one, where the object is to distribute the pressure upon many points; but as an abutment can always be made sufficiently strong to resist the thrust of a stone arch, it cannot be supposed that there would be any difficulty in guarding against the pressure of a much lighter structure of cast-iron. Certain it is, that all the arches cannot act at the same distance from the neutral axis, and therefore a smaller quantity of material at the maximum distance would be equally efficient. No new principle is involved in the construction of iron bridges; the strains are disposed, and must be guarded against, in the same way as in wooden structures; the only modifications are those required by the peculiar character of the material, and by the greater difficulty of securing proper connections.

Fig. 72.

$A\,B$ represents an arch of iron, constructed of plates of considerable lengths, laid upon each other so as to break joint, and bolted together, or in any other suitable way, $C\,o$ being the greatest rise that can be given to the arch.

($s\,s'$) are vertical posts or columns, which may be either of cast-iron or wood: the latter would better resist an impulsive force, owing to its superior elasticity; but the former would be more elegant, and by using a deep string-piece of wood ($m\,m'$) as a cap, there would probably be no danger of fracture from the impulsive force of a passing locomotive. In fact, whatever may be the plan of the structure, it would not be proper to place the rails of a railway immediately upon cast iron supports; there should be some elastic substance interposed to break the force of shocks.

It is probable, therefore, that no objection could be made to the use of hollow columns for the posts $s\,s'$.

$A\,s$, $r\,s'$ are rods of wrought-iron, with nuts and screws, designed to counter-brace the arch, on principles already explained, and prevent it from rising by the action of a heavy variable load upon the opposite side or at the centre.

If the depth of the arch is not sufficient to prevent the centre from rising, by the action of loads upon the haunches, the roadway ($m\,m'$) may be raised so as to admit of diagonal ties between it and the arch at C; or, in some cases, the handrail $n\,n'$ may be so connected with the truss as to form a very efficient counter-brace, or a slight inverted arch could connect the interval $r\,r$, which might be placed below, or if placed above $r\,r$, it might form part of the railing of the roadway; or, lastly, a straight or arched piece could connect $p'\,p$, and the system of diagonal ties be continued to the centre.

Such a combination is perhaps the lightest that could possibly be made to span a given interval.

We will now examine the effect of expansion.

The effect of expansion will evidently be to cause the arch to rise, or to increase the versed sine $o\,C$. This rise will diminish towards the abutments, where it becomes nothing; but the greatest strain upon any of the connecting rods will be at the first quadrilateral $A\,s$, because here the angular change of figure will be greatest. The effect of the change is to strain the tie $A\,s$, but to compensate for this is the expansion of the tie itself, by the same change of temperature which affects the arch and the elasticity of the beam $m\,m'$, which will stretch and yield to the strain caused by the extension of the tie.

The greatest extension of plate or bar iron, when exposed to the extreme variations of atmospheric temperature, is ·001 of its length according to experiments made by the writer, and the extensibility of wood according to Tredgold, is $\frac{1}{470}$ of its length, without injury. From these data, the relative extensions in any given case can be calculated. In an arch of 500 feet span, and 50 feet rise, the extension amounts to $\frac{5}{10}$ of a foot or 6 inches.

The effect of this expansion, if the truss is of the form represented in the diagram, is more than half counteracted by the expansion of the ties in the most unfavorable case, and when the posts which support the roadway are not very far apart, the expansion of the ties may, of itself, be more than sufficient to counteract the expansion of the arch; but even if it should not be, the only effect would be to extend and compress laterally the wooden beam $m\,m'$, which is able to bear without injury four times the extension which change of temperature would produce upon the arch. It is reasonable to suppose, then, that a system connected in this way would have nothing to fear from changes of temperature.

Other forms of trusses are more liable to be affected by changes of temperature, and it is important in arranging the details of an iron truss, to take this fact into consideration; the extension of bar iron within the elastic limits, is as great as that caused by atmospheric changes, and this elasticity is in general sufficient to effect a compensation, and prevent any injury from excessive strains. The principle of the counter-braced arch seems to be peculiarly well adapted to the construction of iron bridges, as the compensation is almost perfect, and the only effect of expansion or contraction will be, to raise or depress very slightly the crown of the arch.

Arches composed entirely of cast-iron have been much used for bridges in England, but the author does not place much confidence in the material, where it is liable to be subjected to impulsive forces; an arrangement, which he considers far preferable, and which has been adopted for two of the bridges on the Pennsylvania Railroad, consists of rolled plates laid one upon another so as to break joint, and clamped together, with or without a centre rib of cast-iron.

APPLICATION OF RESULTS.

THE results deducible from the preceding general theory of Bridges, will now be condensed and practically applied, to determine the proportions of **the parts of a bridge of assumed** dimensions.

In proportioning the parts of structures it **is customary, and** also highly expedient, to throw a considerable excess of strength in favor of stability, and many practical men **have** even repudiated theory altogether, as leading to results which cannot **be relied upon.** The **fault has** been, either that the theory **itself was erroneous, or** sufficient allowance was not **made** for imperfections of material and workmanship.

With a theory confirmed by experience, and with resisting **powers** assigned **to the** materials, sufficiently far below the **limits given** by experiments on perfect **specimens,** the utmost confidence can **be** placed upon the results.

Dimensions arbitrarily assumed, **in** accordance **with the** usual custom, are certainly less to be relied upon **than those** determined upon correct **principles of** calculation.

In determining **the weights of** bridges, it is necessary to prepare a bill of timber from assumed dimensions, and multi**ply the number of** cubic feet, by the weight per cubic feet of **the material;** which we will take, as an average, at 35 pounds. **The quantity of timber** will be assumed (in the following calculations) at 30 cubic feet per foot lineal, as this is about an average of the Howe bridges, on the Pennsylvania Railroad. The greatest load that can ever be thrown upon a railroad bridge, would consist of several locomotives, of the first class, attached together—as is sometimes done in clearing off snow in winter.

The heaviest locomotives in use weigh about 23 tons, and their length is 23 feet. Consequently, 1 ton per foot for the load, and ½ ton per foot for the weight of the structure, may be assumed as a safe average for the maximum load, where the span does not exceed 200 feet. One and a-half tons per foot lineal, will, therefore, be assumed as the extreme load— in the following calculations.

The greatest safe strain per square inch for wood, will be considered as 1,000 pounds, and for iron, as 10,000 pounds.

To determine the strain upon the chords.

The strain upon the upper chords, is one of compression; it is greatest in the middle of the bridge, and diminishes towards the ends. The maximum strain in the middle, is equal to that force which, if applied horizontally, would sustain one-half the bridge, if the other half were supposed to be removed. To obtain it, multiply half the weight of the bridge by the distance of the centre of gravity from the abutment (which is always very nearly one-fourth the span), and divide the product by the height of the truss, as measured from the middle of one chord to the middle point of the other.

Let H represent the horizontal strain in the centre
S " " span of the bridge
h " " height of truss, from middle points of chords
W " " weight of the whole span

Then $H = \dfrac{W}{2} \times \dfrac{S}{4} \times \dfrac{1}{h} = \dfrac{SW}{8h}$.

Example.

If the span of a bridge be 160 feet, and the height of truss 17 feet, what should be the cross-section of the upper chord in the centre?

The weight, calculated at 1½ tons per foot lineal, will be 480,000 pounds. If the chords be 12 inches deep, and 17 feet be taken as the measurement, from out to out, the distance from the middle of upper to the middle of lower chord will be 16 feet; applying the formula, we have

$$H = \frac{480,000 \times 160}{8 \times 16} = 600,000 \text{ pounds.}$$

The cross-section of the chord, to resist this compression, at 1000 pounds per square inch, will be 600 square inches; and as the depth is 12 inches, the total breadth must be 50 inches; or, 25 inches to each truss, if there are two trusses.

Of the strain upon the lower chord, at the centre.

The strain on the lower chord is equal in degree to that of the upper, but it is tensile, while the former is compressive.

If the chord could be made in a continuous piece, without joints, the dimensions would not be required greater than in the former case; but, as there is generally one joint in every panel, it becomes necessary to increase the quantity of material to such an extent, that the resisting area, exclusive of the joint, shall be sufficient to resist the strain.

This requires, in general, that one additional line of chord timbers should be introduced. It is a good practical rule (and one which is observed in Howe's bridges), to make the upper chord consist of three, and the lower of four timbers to each truss; a joint will then occur in each panel, and the pieces should be sufficiently long to extend over four panels. With this arrangement, three of the timbers must be allowed to sustain the whole strain, since that which contains the joint is not capable of opposing any resistance.

Strain at the ends of the chords.

In a beam resting on two supports, the strain at the end is nothing, and increases uniformly to the centre, but in a bridge-truss of a single span, there will be a horizontal strain at the end of the brace, nearest the abutment, which will equal the weight on the brace multiplied by the co-tangent of the inclination of the brace. **If the inclination of the brace is 45°, the** horizontal strain will be equal to the vertical weight **upon** A. If (as is generally the case) the angle with the horizontal is greater than **45°, the** horizontal strain will be less than the weight, **and** consequently, it will be safe **in** practice to assume the horizontal **strain at** the **end of the chord, or more correctly** at the end of the first brace, as equal to the vertical force acting on **that brace.**

This vertical force is one-half the whole weight of the bridge, and if we continue the calculation with the dimensions already given, the half weight will be 240,000 pounds, and the cross-section to resist it 240 square inches, or a little more than one-third the size, required in the centre.

Having determined the cross-section of the chords at the centre **and end, a** uniform increase between these points will fulfil all the necessary conditions.

Another circumstance must be taken into **consideration, in** determining the size of the chords. The applied weight produces a cross-strain upon that portion of the chord which lies between any two posts, and the condition of the chord is that of a beam supported at the ends, and loaded in the middle. The formula is $R = \dfrac{3\,w\,l}{2\,b\,d^2}$, or, as ($b$) is the quantity to be determined $b = \dfrac{3\,w\,l}{2\,d^2\,R}$. If the interval of one panel be assumed **as** 12 feet, (l) expressed in inches, will be 72, $d = 12$, $R = 1,000$, $w =$ weight in an interval of 12 feet, which cannot exceed 6 tons applied at the centre. By substitution, we obtain

$$b = \dfrac{3 \times 12{,}000 \times 72}{2 \times 1{,}000 \times 12^2} = 9 \text{ inches.}$$

Hence it appears that the size of the chord, as determined by this condition, is much less than by the former, and consequently, the dimensions previously given are ample to resist the cross-strain arising from the passing load.

*Of **the** strain upon the ties and braces.*

These strains are always estimated together, because they bear to each other at every point, the proportion of the height of a panel of the truss to the diagonal; so that if one is known, the other is readily determined by a simple proportion.

In a simple truss, consisting of chords, ties, and braces, the braces which project from the abutments sustain the whole load. The weight is not distributed equally amongst all the braces, as one unacquainted with the action of the system might suppose. The proportional strains on each successive brace, from the centre to the ends, may be illustrated by a chain suspended from a fixed point; the upper link sustains the whole weight, the lower none; each link transmits the weight of those below to the one above it; and similarly, each brace transmits the strain from the middle of the span to the end, adding to it the portion due to the panel of which it forms a part. The end braces, unless relieved by an arch, sustain the whole weight of the structure, and its load.

As the weight of the bridge under consideration is 480,000 pounds, each end must sustain 240,000 pounds, and at 1,000 pounds per square inch; in the cross-section, the ties must be 240 square inches, if of wood, and 24 square inches, if of iron, allowing in the latter case 10,000 pounds per square inch as a safe load, although in practice it is sometimes greatly exceeded.

If the panels be 12 feet wide, and 16 feet high in the clear, the diagonal or brace will be 20 feet, and the strain on the brace will be $\dfrac{240,000 \times 20}{16} = 300,000$; thus requiring 300 square inches of wood, or 4 braces, of 75 square inches each.

The strain, exactly at the middle point with a uniform load, is theoretically nothing; and it increases from this point to the end, where it attains a maximum equal to one-half the whole weight of the bridge; but in practice there never is a brace exactly at the middle; the panels must have considerable magnitude in a horizontal direction, and the proper estimate of the strain is that which would be produced by half the maximum weight on two adjacent panels, or the whole weight on one panel. This weight, in an interval of 12 feet, will be 18 tons, or 36,000 pounds; requiring a cross-section of only 36 inches, or 9 inches to a tie, and $11\frac{1}{4}$ to a brace.

In this case, the cross-section of the brace, as given by the condition that the strain shall not exceed 1,000 pounds per square inch, is too small, since its length would cause it to yield by lateral flexure. To obtain the proper dimensions for the brace, in this case, we must have recourse to the formula for long posts; which, for white pine, gives $w = \dfrac{9{,}000\, b\, d^3}{l^2}$.

If we assume that the brace is 20 feet long, without intermediate support, and that the depth is 5 inches, the breadth will be, $b = \dfrac{w\, l^2}{9{,}000\, d^3} = 4$.

The condition which has been assumed is not a common one; the braces are almost always supported in the middle, which reduces the length of the unsupported portion to one-half. In this case, the formula would give one-fourth the breadth required in the former case. Very small braces would, therefore, be sufficient at the middle of the bridge, if supported at the middle of their length.

It was calculated that each of the four braces in the end panels required 75 square inches of cross-section to resist the strain, allowing 1,000 pounds per square inch. It is possible that, even with dimensions sufficient to furnish this area, they may yield by lateral flexure. To test this, assume the depth as 9 inches, and breadth $8\frac{1}{2}$ inches, and substitute in the expression for the weight, which becomes $w = 185{,}880$ pounds; a result which proves that the flexure is impossible with the weight and dimensions assumed, and that the pressure in the direction of the brace is the only one to be provided for.

Having determined the dimensions of the ties and braces, at the centre and ends of the truss, the intermediate timbers should increase by regular additions. If the truss contain 16 panels, the section of the middle braces containing 15 square inches, and the extreme brace 75 inches, the intermediate braces should be in regular proportion, thus: 15, $23\frac{1}{4}$, $32\frac{1}{4}$, $40\frac{3}{7}$, $49\frac{3}{7}$, $57\frac{6}{7}$, $66\frac{3}{7}$, 75.

In a truss thus proportioned, the middle braces contain only one-fifth the material of the end braces, and the ties should be in the same proportion.

Counter-braces.

The conclusion arrived at, in considering the subject of counter-braces, was, that the greatest strain upon a counter-brace was equivalent to that produced by the action of the greatest variable load upon a brace; it will consequently be equal to the strain upon the braces of the middle panel; and if each panel contains two braces, and one counter-brace, the size of the latter should be uniform, and equal to 30 square inches of cross-section, in the truss under consideration. A counter-brace, 5 × 6 inches, supported in the middle, would afford the requisite proportion.

Lateral horizontal braces.

The use of lateral bracing is principally to guard against the effects of wind, and other disturbing causes, tending to produce lateral flexure in the roadway. The ordinary bracing to resist this action consists of ties and braces, similarly disposed to those in the main truss, except that equal strength is required in the direction of each diagonal of the horizontal panels. The greatest lateral strain is that produced by the action of a high wind; assuming the force of wind at 15 pounds per square foot, as a maximum, and allowing the height of truss to be 18 feet, the uniform weight over the surface, if weather-boarded, would be 43,200 pounds, or 21,600 pounds to each series of braces, top and bottom. The effect of this force would be estimated, precisely as the strain of a uniform load upon a bridge, and if the angles of the lateral braces be 45°, the diagonal would be to the side as 1. 4 : 1, nearly. The strains on the end braces will be, $\frac{21,600}{2} \times 1.4$ = 15. 120; which, at 1,000 pounds per square inch, would require but $15\frac{12}{100}$ square inches to resist the strain; or, if the brace is 16 feet long, supported in the middle, and depth 5 inches, the breadth as determined by the formula $b = \frac{w\,l^2}{9,000\,d^3}$ will only be $1\frac{3}{10}$ nearly: 4 × 5 would, therefore, be sufficient for the largest lateral brace at the ends. At the middle of the span the lateral braces would be exceedingly light; they might even be omitted in the middle panel without injury.

In long spans, this diminution in the sizes of the braces in the middle adds considerably to the strength, by relieving the bridge of unnecessary weight.

Diagonal braces.

These timbers occupy the direction of the diagonals of the cross-section of the bridge; they are admissible only when the roadway is on the top, and are of great utility in preventing side motions; where the roadway is on the bottom, knee-braces must be substituted.

The strains upon these timbers, which result from unequal settling, cannot be calculated, as it is impossible to determine the side to which the bridge may have a tendency to settle; the only rule is to make them large enough.

Experience has shown that 5×6 is sufficient for knee-braces, and 5×7 for diagonal braces in ordinary cases.

Floor beams.

Allowing the unsupported interval between the trusses to be 14 feet, the depth of beam 14 inches, and the greatest load equivalent to 6 tons, applied at the centre; required the breadth to allow a deflection of $\frac{1}{40}$th inch to one foot: timber, white pine. The formula for this case, is $w = \frac{B D^3}{\cdot 0125 \, l^2}$ whence, $B = \frac{w \, l^2 \times \cdot 0125}{D^3} = \frac{12,000 \times 14^2 \times \cdot 0125}{14^3} = 10\frac{5}{7}$ inches.

Having now completed the calculations for all the parts of an ordinary truss, composed of chords, ties, braces and counter-braces, it remains to estimate the effects of the introduction of arches, and the combinations of different systems with each other.

In entering upon the consideration of this subject, it is proper to remark, that our calculations must be based to some extent upon uncertain data; for where two systems are combined, we cannot be certain that each sustains an equal portion of the weight; but, on the other hand, we are sure that the assertion sometimes made, that either one or the other necessarily sustains the whole load, is erroneous. Much depends upon the manner of making the connection; if an ordinary truss be constructed, and arches added after it has settled to a considerable extent, by the application of heavy weights, it is very clear that the arch will bear but a small proportion of the load; but if the arch is introduced, previous to the removal of the false works, and both systems be allowed to settle together, it is fair to suppose that the strain upon each will be in proportion to the respective power of resistance.

The usual method of constructing bridges, is to make the truss of such strength as is supposed sufficient to support the weight, and to add the arch as additional security. We think it decidedly preferable to reverse this arrangement, making the arch the main dependence, and using a light truss in combination with it, merely to prevent change of figure in the arch, and to give the proper elevation or inclination to the roadway.

Let a railroad bridge of 160 feet span be supported by four arches, the rise of each of which is 20 feet; weight on bridge, $1\frac{1}{2}$ tons per lineal foot—required the dimensions of the arches to sustain the whole weight.

The whole weight is 240 tons, or 240,000 pounds to each half of the bridge. The strain upon the arches in the centre will therefore be $\dfrac{240,000 \times 40}{20} = 480,000$ pounds, requiring 480 square inches of cross-section, at 1,000 pounds per square inch.

Four arches, 16 inches deep and $7\frac{1}{2}$ inches wide, could supply the requisite amount of material. The compression at the ends will be to that in the centre as $\sqrt{40^2 + 20^2} : 40$, or as $\sqrt{2^2 + 1} : 2$; hence, it will be $480,000 \times 1\frac{1}{8}$ nearly, $= 540,000$, and will require 540 square inches; or, if the arches are $7\frac{1}{2}$ inches wide, as before, the depth must be 18 inches.

If an arch is too small to sustain the whole weight, and is connected with a truss of given dimensions, the best practical manner of treating the case is to estimate separately what each would sustain, allowing 1,000 pounds per square inch;

and divide the weights in proportion to the powers of resistance. Having, in this way, determined the weight to be sustained by the truss, the parts can be proportioned in the manner previously explained.

It is very evident that an arch can be made to sustain the whole of the weight, for if a truss has settled it may be raised to any extent, by the addition of arches and suspension rods. In this case, the principle of proportioning the braces, so as to increase in arithmetical progression from the middle to the end, is no longer applicable; there is no more strain at the ends than in the centre, and but little at any point, and in this case the truss is of no other use than to stiffen the arch and carry the roadway.

Amount of counter-bracing which an arch requires.

That a very slight force is sufficient to counter-brace an arch, may be rendered evident, without a calculation in detail, by taking a more unfavorable case than could possibly occur in practice. Let A and B (Fig. 97) represent the skew backs of an arch, and leaving out of consideration the resistance of the lower chord, which adds greatly to the stiffness; suppose a weight of $1\frac{1}{2}$ tons per foot to be placed on one-half of the arch, the weight of the other half, being $\frac{1}{2}$ ton per foot, will leave 1 ton per foot to produce a change of figure. The effect of this weight will be represented, nearly, by one-half applied at the middle point (p). Let the span, $s = 160$ feet; and the rise of the arch, $r = 20$ feet, the weight at $p = (w)$ will be 40 tons. Let H represent the component of the weight, in the direction of the chord $A\,p$. At the centre, the value of this component would be $\dfrac{w}{4\,r}\sqrt{4\,r^2 + s^2} = 82$ tons, and as it is always less at every other point, the slight error will be on the safe side, by taking 82 tons as the force acting along $A\,p$. This force, and its equal at A, gives a resultant, acting upwards at m, which is expressed by $\dfrac{o\,n}{o\,p} \times 82 \times 2$. In the present example, $o\,n = 12$ and $o\,p = 60$ nearly; hence, the force at m, which acting upward must be resisted by the counter-braces, is 65½ tons, requiring only 65½ square inches of resisting area to each side truss. A single stick, 8 × 8, would therefore be nearly sufficient, and when it is considered that the strain is not all at one point, m, as we have supposed it, but is distributed over a considerable length of arc, the amount of counterbracing necessary to resist it must be very small.

The principle of determining the size of the counter-brace, by the force that would be required to resist the upward action of the arch, is not that which we recommended when this subject was considered.

It is preferable to make them sufficiently strong to throw a permanent strain upon the arch equal to that produced by the passage of the load, and this condition requires as much resisting surface as that presented by the middle braces. It is unnecessary to continue the application of these principles to a greater extent; we believe that every case of much practical importance has been considered; and the illustrations given will be sufficient to indicate the manner in which the results obtained can be applied to the determination of the dimensions of other structures. We propose, in the second part of this work, which will be devoted to an examination of particular modes of construction, to enter more into detail, when an opportunity will be offered of supplying any deficiencies that may exist, and of illustrating the modes of calculation by which the strains may be determined, and the parts proportioned, in every variety of combination.

EQUILIBRIUM OF ARCHES.

WE cannot, perhaps, introduce this subject better, or express our own views of it more clearly, than by presenting the reader with the following brief exposition of the ordinary mode of investigation, as copied from a manuscript procured from a brother engineer. It exhibits no new principle, and the formulas are deduced in a similar manner to that which has been used in Hutton's Mathematics; but as the facts which it contains are important, and must form the basis of every correct theory for determining the conditions of equilibrium, we will give the explanation of this method, and then proceed to point out what we believe to be its defects.

PROBLEM.

To find the thickness of abutments of arches, of any kind.

From Gauthey (modified).

"It has been found by Monsieur Boistard, who built some good bridges, that there were certain points in an arch that were weaker than others, which give way at the moment when it fails. These points are denominated by him the points of rupture, and are very necessary to a proper solution of this interesting problem, which is now very much simplified by the author above named." Gauthey took up the experiments of Boistard, and upon them has founded the following solution.

Case 1*st.*

FIG. 73.

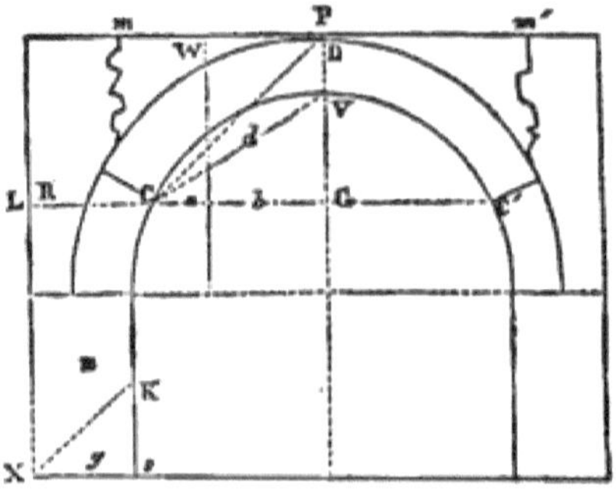

Let CVC' be the intrados of any arch, whether semicircular, elliptical, Gothic, or composite. Let D be the crown of the extrados, or back of the arch, which is supposed to be filled up level with the haunches at m and m'. If a weight be placed upon the crown too great for it to bear, it yields, and the arch-stones open beneath, at the crown, while the extrados is found to open at some point on each side; either at the spring, if it be a flat arc of a circle, or about 30 degrees of a semicircle, or at various other points if it be composed of arcs of circles, tangent to each other, and of various rises, whether $\frac{1}{4}$, or $\frac{1}{3}$ or $\frac{1}{5}$ of the span, and the arch only falls by pushing aside the abutments at C and C', the opening at R extending itself up to the top at m and m'. The parts of the arch comprehended between the joints of rupture are called *acting*, and the rest *resisting*. It has, moreover, been observed that when the abutment gives way, it leaves a portion of itself standing, viz., XKs; the line XK being at an angle of 45° with the horizon, which only adheres by the strength of the mortar or cement made use of.

These facts being stated as above, we may now consider the manner in which the upper part acts to overturn the abutments, and how they resist that action.

Let the weight of the portion $CRmP$, on one side of the crown, be represented by w, this weight may be conceived as supported by two points C and D and pressing upon them inversely, as their distance from the vertical line passing through the centre of gravity of that portion. If that vertical cut CG at Q, and we call CQ, a, and CG, b, then the whole weight is to the part resting on C, as $b : b-a$, and the whole weight to the pressure at D, as $b : a$. Now the pressure at C acts merely to keep down the abutment, and that by a leverage LC; but the pressure at D produces a different effect, and one that must be carefully attended to. Draw CD and DG, and call DG, c, and CD, d, for brevity. The two half-arches press together at D, and mutually sustain each other. Let the pressure on one side be represented by DG (c), and let it be considered separately, and apart from the other side, (c) may be decomposed into an oblique thrust (d), and a horizontal action (b), which last acts towards G, and tends to crush the stones at the key, and is met and resisted by the strength of the stone, strongly confined between the pressure (b) and its equal and opposite pressure (b'), so that we have only to consider the oblique action (d) which evidently bears from (D) towards (C), and partly tends to press (C) horizontally, and partly to keep it down vertically, and this is to be added in part to the resisting forces; and in proportion as (Cd) is more nearly horizontal, so much the more powerfully it presses (C) horizontally, and *vice-versa;* as it is more vertical the more does it tend, as in Gothic arches, to weigh down the abutments and keep them steady. It is, moreover, the oblique pressure which this part of the arch exercises, which squeezes the arch-stones so tight together between the crown (D) and the point of rupture (C), as to make them act as one homogeneous mass, or stone, whose individual parts cannot slip out, even though they should not be shaped as wedges.

The former notion, about the arch being perfectly equilibriated by a catenarian curve, is now regarded as a fallacy,*

---

* We agree with the writer, when the catenarian curve is taken as the intrados, but when it is used to determine the direction of the joints, and the latter are made perpendicular to it, we regard it as any thing but a fallacy. With as much propriety might the practice of building a vertical wall with horizontal courses, that is, with beds perpendicular to the line of direction of the pressures, be regarded as a fallacy

and the whole matter at present rests upon the relative degrees of action of the upper and lower parts, or the parts above and below the points of fracture. It may be proper here to remark, that any additional weight upon the crown, such as **is often** seen in heavy banks over culverts, may be easily taken into consideration, as all that part which rests vertically over the acting portion, tends, through their common centre of gravity, to produce similar results to the masonry itself, and all the additional weight, which is just over the resisting parts, has the effect of keeping the abutment in its place. We should **not** regard the pressure of earth behind the walls (although this has undoubtedly a very great effect in preserving their stability), because, by some flood of the stream or canal the embankment may be washed away, and then if the abutments had not been calculated to sustain the pressure of the arch, they will be overthrown. Besides, earth is very compressible, and the abutments, although sustained from actually falling by the pressure of the bank behind them, may yield a little, and thus disfigure the work.

Having thus premised the general considerations, **we are** now prepared to go into the algebraic forms for expressing **the** exact quantities of each of the acting and resisting forces.

First.—For the portion of (w), resting on (C) and (D),

$$b : a :: w : \frac{w\,a}{b}, \text{ the weight resting on } D,$$

$$b : b - a :: w : \frac{w}{b}(b-a), \text{ the weight resting on } C-(A).$$

Let the weight on D be decomposed into the oblique force acting in (d), and the **horizontal one in** (b), thus,

$$c : d :: \frac{w\,a}{b} : \frac{w\,a\,d}{b\,c}, \text{ for the oblique force acting from } D \text{ towards } C,$$

$$c : b :: \frac{w\,a}{b} : \frac{w\,a\,d}{b\,c}, \text{ for the horizontal crushing force, acting}$$

from C towards G. This last is neutralized by its opposite, $C'\,G$.

The oblique force from D towards C, is now to be decomposed into two others, **the one** acting downwards at C, the **other acting horizontally from** G towards C.

$d : c$ as $\dfrac{w\,a\,d}{2\,b\,c} : \dfrac{w\,a}{b}$ which is nothing more than to say that the weight on D is transferred by the oblique action to C, for it is the very same as the first expression above for the weight on D; add, therefore, the last obtained vertical force on C to the second expression (A), showing the weight on C, and we have

$$\frac{W\,a}{b} + \frac{W}{b}(b-a) = \frac{w}{b}(b+a-a) = \frac{w\,b}{b} = W$$

showing that however the forces have been considered under their various actions, they still result in simply resting one-half of the upper portion of the whole arch on C, the other half being borne by C', which is obvious. Next, to obtain the horizontal thrust produced by said oblique force acting in $D\,C$,

$$d : b :: \frac{w\,a\,d}{b\,c} : \frac{w\,a}{c}$$

This last is the only force, which has any effect to overturn the arch, and acts by its leverage $L\,X$, which we may call (e), X being the fulcrum or pivot around which it would turn in its overthrow; $\dfrac{w\,a\,e}{c}$ is then the moment of the acting force.

W, the weight of half the upper part, acts at C at the end of the lever $X\,S$ which we shall call (f), to keep the arch in place. If (u) be the weight of all the resisting portions, which may properly be called abutments, and a vertical be passed through its centre of gravity, falling at a distance (g) from X, then (u) has (g) for its lever, and ($u\,g$) is the effect of the abutment to resist the action, to which we must add $w\,f$, for the sum of all the resisting forces, thus, $u\,g + w\,f$.

It is needless to say that $u\,g + w\,f = \dfrac{w\,a\,e}{c}$ when in equilibrium. If the lower part is built with the best hydraulic cement, it may be, that its cohesive force on the joints of fracture $X\,K$, will keep the triangle $X\,K\,S$ from separating in its overthrow; this will evidently depend on the quality of the cement, and would lead us to conclude that no expense should be spared to make the mortar of the abutments as good as possible in all cases.

It may be useful here to adduce by way of practical illustration of the above theory, a case which did actually occur, in the year 1829 or '30, in constructing the Chesapeake and Ohio Canal. The Monocacy, a very violent stream, is crossed by a beautiful stone bridge (aqueduct), of 9 arches, each 54 feet span, and 9 feet rise; arches $2\frac{1}{2}$ feet thick, abutments 10 feet thick, and 10 feet high on a foundation of 3 feet high, and 13 feet wide.

Some arches and piers had been built up and backed in, but, before the whole could be completed, a great flood swept away the last centre from under the arch just turned and not backed in, except partially on one side: the rise of this arch being only one-sixth part of the span, must have pressed with tremendous effect upon its last pier, especially as the supports were very suddenly knocked from beneath it, and it was brought to bear very suddenly upon the pier. This had been well built with hydraulic cement of tolerably good quality, only eight or ten months before. The arch stood triumphantly, and contrary to the expectation of all that witnessed it, who looked for nothing but the destruction of every arch then built one after another. But, upon a subsequent investigation of the thrust and resistance,

the former $\dfrac{Wae}{c}$ was found to be 126,300 pounds,

the latter $ug + wf$ was found to be 162,390 pounds, showing a considerable surplus of strength. Its standing was then attributed to the great strength of the cement, and excellence of workmanship; but the cement having been since tried, and found to slake in the air like common lime, it was no better than good lime-masonry should always be, and its standing must be attributed to intrinsic weight and strength.

$a = 12.5$ $g = 5.$
$b = 27.$ $w = 11,620$ pounds
$c = 11.5$ $u = 9,438$ weight of a cubic foot of stone, assumed $= 140$ pounds
$e = 10.$
$f = 10.$ w having lost much of its specific gravity by immersion.

Case 2d.

There is another point of view, under which the yielding of an arch may be considered. The upper portion acting as above mentioned, and the lower parts, or abutment, being sufficient to resist them and to keep from being overturned, the arch may give way from the breaking up and sliding of the stones one upon another, at some horizontal joint; suppose for instance, at LC: in that case, the resistance must depend entirely on friction, and on the strength of the mortar.

Boistard has found, from numerous experiments on the subject, that friction is $0.76\,w$ (w being the weight resting over the joints of rupture), and that the strength of adhesion of the mortar is 3,900 pounds per square foot. Now assuming this as true for a majority of cases, though evidently subject to much modification, and dispensing with the leverage used in the first case, we shall have the horizontal thrust simply $\frac{Wa}{c}$. And calling the strength of mortar per square foot (s), and the number of square feet area of mortar joint (h), (u) the superincumbent weight above the joint, and (r) the friction of the sliding parts, the resistance will be $(W+u)r + sh$; and when on exact balance, $\frac{Wa}{c} = (W+u)r + sh$.

That is, the sliding joint (assuming several for supposition), where the resistance is found least in relation to the sliding force opposed to it, is generally at the springing line, for above that the area of joint increases rapidly, and below it the weight causes more friction.

This leads to the practical consideration which has made eminent bridge builders change their former practice very much, that the more uneven and projecting the stones in the abutments, near the springing line, and the more inclined towards the thrusting line, the more effectually will they resist. For this reason arches are sometimes continued through the abutments; at other times, stones are set up at frequent intervals on end, amongst the others, and the masons are forbidden to course behind.

Gauthy found that, in some cases compared by him, it requires a greater thickness of abutment for the second case than for the first.

	Thickness of abutments.	Pos. of joint of fracture.	Thickness of abutments.	Pos. of joint of fracture.
For semicircles	1·47 ft.	30°	4·31 ft.	15°.
anse-paniers rising ½ the span	2·16 ft.	50°	5·30 ft.	35°.
anse-paniers rising ¼ the span	2·68 ft.	60°	7·32 ft.	45°.

The arch above supposed, is 67·4 feet span, arch-stones 3·27 feet long, backed up level, and springing from the broad platform of the foundation without any height of abutment.

But where the abutments have height, as in ordinary cases of smaller arches, the thickness found by him would have been vastly increased, on account of the great increase of thrust from greater leverage.

The actual position of joints of fracture can only be found by trial of several suppositions, and that is to be taken where the resistance is weakest when compared to the thrust at that point, or where they are most nearly equal, and consequently their ratio is the least.

It is well to remember that the resistance is much diminished when the abutments are immersed in water, as in piers in rivers.

In the case of the Monocacy aqueduct, tried upon the last supposition, we find $\dfrac{w\,a}{c} = 12{,}630$ pounds, and $(W + u)\,r + s\,h = 50{\cdot}023$ pounds, showing a greater excess of resistance than in the other case, supposing it to yield by overturning.

If in an arch the joints of rupture be at the springing lines and the extrados of the crown, the horizontal thrust is $\dfrac{a\,w}{b}$, in which a is the distance from the springing line to the perpendicular, through the centre of gravity; b is the vertical distance equal to the rise of arch + thickness of ringstones, w the weight of half the arch.

The conditions of equilibrium of the abutment are simply that the moments of the horizontal and vertical forces shall be equal, the weight of the arch being applied at the springing line.

APPLICATION OF RESULTS.

Although it would appear from the preceding statements, copied literally from the manuscript referred to, that the results given by this formula can be relied upon in practice, yet, notwithstanding the evidence furnished by the Monocacy aqueduct, we cannot think that the dimensions given by the formula are sufficient. When the stone is extremely hard, and the pressure upon it very small in proportion to its capability of resistance, the result may be sufficiently great, but in other cases it cannot be trusted. In fact, it is evident that the formula has been deduced upon the supposition that the pressures are thrown entirely upon the points D and C, but, unless the strength of the material be almost infinite, these points could not sustain the pressure; the portions of the stone lying at these points would break off, and the points of contact D and C, being thus brought nearer together, would render the line of direction of the pressure more nearly horizontal, increase both the horizontal force at C and the leverage $c\ s$, or $L\ X$ at which it acts, and consequently require a greater thickness of abutment to resist its effects.

That which we believe to be the true method of determining the equation of equilibrium of an arch, can be deduced from a process of reasoning analogous to that employed in the case of a straight beam supported at the ends, or the chords of a straight bridge.

The lower fibres of a beam, and the lower chords of a straight bridge-truss are in a state of extension, and the upper ones of compression, and the neutral axis is, in general, in the middle of the depth; but in an arch of any material, resting upon fixed abutments, the resistance of the abutments exactly replaces that of the ties or lower chords in the former case, and the position of the neutral axis will remain unchanged.

Fig. 74.

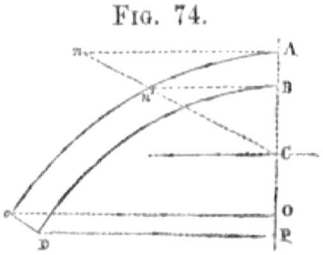

Let DB represent a half arch. Draw AP, Oo' and DP. If $OP = AB$ the resistance of the abutments acting in the direction DP will produce the same effect as a tie in the same direction, and capable of opposing the same resistance. Since, therefore, there is a change from extension at P, to compression at A, there must exist, as in beams or straight bridges, a neutral axis between A and P; and as AB, as will be shown, equals OP, the neutral axis will bisect AP.

The pressure upon any given point of the joint AB, will be as its distance from the neutral axis; and if the perpendicular An represents the maximum strain upon a square unit at A, join Cn, and the perpendicular of the triangle ACn will represent the proportional pressures upon other points. The whole pressure upon the joint will be represented by the trapezoid, Bn. A perpendicular to AB, through the centre of gravity of the trapezoid, will give the centre of pressure of the joint AB, which, when CB equals or exceeds AB, or in other words, when the rise of the arch is greater than about **three or** four times the depth of the arch-stones, will be sufficiently near the centre of the joint to render the error made by taking it at the centre very small, and that too on the **side of** stability.

When greater accuracy is required, the centre of gravity of **the** trapezoid must be found. As a general rule, we think that practical formulas of this kind should be made as simple as possible, and that instead of aiming at the greatest theoretical accuracy, it is best to reject small errors that are in favor of stability, in order that the formula may give an excess of strength. As an illustration of the little reliance that practical men place upon the deductions of theory, we will state, that the dimensions assigned to parts of structures are often twice **as** great as **the rule allows. Such a difference should** not **exist; the dimensions** of structures deduced from theoretical **considerations should** correspond with those assigned in practice, **and in order that** this may be the case, the theory must **be based on** correct principles, and include every circumstance **which tends to** derange **the** stability.

The ordinary equations of equilibrium will **therefore give**

results sufficiently near the truth, by taking the middle of the joints, instead of the points A and D, as the points of application of the pressures.

It is very necessary to observe that the equation of equilibrium above determined is based upon such conditions, that the resultant of all the forces, both of the acting and resisting portions, passes through the point x at the back of the abutment. The dimensions thus determined will be sufficient only in the case of a rock, or other incompressible foundations; in other cases, where there is any liability to yield, the resultant, instead of passing through the extremity must pass through the middle of the base. This condition is, in general, best fulfilled by making the back of the abutment in steps or offsets, which permits an enlargement of the base, without greatly increasing the amount of masonry; and, at the same time, favors stability, by throwing the centre of gravity very much towards the face.

If the arch between D and B were in one solid piece without joints, it would follow, that the joints $A\,B$ and $C\,D$, being entirely above and below the neutral axis, would be compressed throughout their whole extent, and would have no tendency to open; but cases have often occurred in which some of the joints have opened at the back or front, and the work suffered considerable derangement. Such an effect may result from two distinct causes.

First.—When an arch is constructed it is usual to commence by laying the stones nearest the abutment, and proceeding towards the centre; months sometimes elapse between the laying of the first and last stones of the arch, during which time, if the cement or mortar is of good quality, those first laid become solidly united to each other. If the centres are removed soon after the completion of the arch, and while some of the joints are in a soft or compressible state, inequalities of settling must result, sufficient in some cases of itself to account for all the observed derangement.

The second case, in which the joints of an arch will have a tendency to open, is when the line of pressure passes below the intrados, or above the extrados. To guard against this effect, the load upon the different **parts of the arch and the curve of its intrados must bear such a relation to each other,** that the line of pressure will never fall outside the limits of any joint, but will approach as nearly to the centre of **the joint as possible.**

To find the relative length of the joints at different points of an arch, **and the line of** *direction of* **the** *pressure.*

FIG. 75.

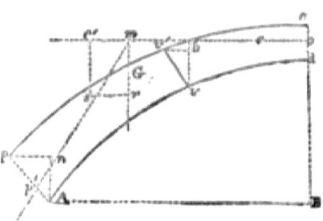

Let $c\,d$ represent the depth of the joint **at the crown necessary** to resist the horizontal **thrust, as determined from assumed** dimensions, and **let** this force be represented by a line $o\,e$, equal to $c\,d$, applied **at the** centre of pressure (o). Let G represent the centre of gravity of the arch $A\,d$, and $m\,r =$ length of line that represents the weight. Transfer the force at o to the point m, and **make $m\,e' = o\,e$.** Construct the parallelogram of forces $m\,s$. As $m\,e'$ represents the length of joint **necessary** to resist the horizontal force, $m\,r$ would be the length **sufficient to** sustain the weight, and the resultant $m\,s$ would **represent** the length of a joint, to resist the combined pressure **of the two forces.** Draw $A\,p$ perpendicular to $m\,s$, produce **and equal in length to $m\,s$.** $A\,p$ will represent both the length **of the joint at the point A, and its** proper direction, since it is perpendicular to the line of pressure $m\,s$.

By drawing $p\,n$ parallel, and $A\,n$ perpendicular to $A\,B$, we find that the triangles $A\,p\,n$ and $m\,s\,r$ will be equal, hence, $A\,n = s\,r = c\,d$, and as the same is true at any other point it follows, that *the difference of level of the extremities* *of any joint of the arch* *should be equal to the depth at the* *crown.* Also as $p\,n = m\,r =$ weight of portion of arch $A\,D$

t follows, that *the horizontal distance between the extremities of any joint will be proportional to the weight of the portion of the arch between it and the crown.* p' being the point of application of the resultant of the pressures upon all parts of the joint $A\,p$, and $p'\,s$ its line of direction, $p'\,s$ must be tangent to the curve of equilibrium. *By finding the point p' for other joints between A and D, the curve traced through them will be the line of direction of the pressures.*

The manner of finding the point p' for any joint $A\,p$ is obvious; it is the intersection of the line $A\,p$ with the diagonal of the rectangle, one of whose sides $e'\,m$ is proportional to the horizontal pressure, and is constant at every point of the arch; the other, $m\,r$, represents the weight of the portion $A\,d$ of the arch, acting through G its centre of gravity. The position of G can be readily found for any joint, as $(u\,u')$ by making a drawing of the arch on pasteboard, cutting it out and balancing the portion, of which the centre of gravity is to be ascertained. The weight can be found either by weighing the pasteboard, or by calculation, and thus we are furnished with an extremely simple and practical method of describing the curve of equilibrium.

The method usually recommended for determining practically the direction of this curve, is to mark off on a wall, or other vertical surface, the span and rise of the arch, then suspend a flexible chain between these points, and load it at short intervals with weights proportional to the superincumbent portions of the arch. As the addition of these weights will change the figure of the curve, the length of the chain and the magnitudes and distribution of the weights must be varied, until by successive trials the proper proportion and distribution are discovered. This, which is recommended as a very simple method, and easy of application by any practical builder, we conceive to be exceedingly troublesome, and such as no practical builder would be likely to undertake; and after the curve has been found in this way, we know nothing of the position of the centres of pressure: in fact, it is evident, from the method which has been pursued, that they have been assumed at the springing lines and at the lowest points of the key-joint, as these are the points through which the curve has been drawn. The method which we have ventured to recommend **determines** the curve at once, without the necessity of successive **trials,** and also gives the centres of pressure of every joint.

In practice, it is not generally necessary to find the curve of equilibrium and trace its course; its principal utility is in determining the direction of the joints, which should be made perpendicular to it, but the direction and length of the joints can be readily determined without it as follows.

We have seen that *the horizontal pressure* **at** *any point of an arch is equal to that at the centre, and is constant;* but the vertical pressure is variable, and equal to the weight of the portion of the arch included between the given joint and the crown; therefore, to find the direction of any joint, as *u u′*, draw a vertical line *u o′* **at the point** *u,* make it equal to *c d,* the depth at the crown, draw *o′ u′* perpendicular to it, and **bearing to it** the same proportion **that the weight of the portion** of the arch *u d* bears to the **horizontal thrust.** Join *u u′*, which will give both the direction and length of the joint.

The above theory, and the simple rule which has been **deduced from it, may be verified to some** extent by applying it to the case of an arch uniformly loaded.

Tredgold and others **have** shown that the curve of equilibrium in **this case** is the common parabola, and the proof is extremely simple.

Fig. 76.

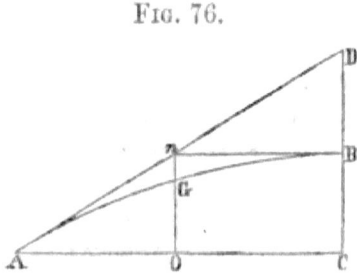

Let *A B* represent a portion of the curve supposed to be **uniformly loaded.** Draw *B n* parallel and *G o* perpendicular to *A C.* **The weight** acting at the centre of gravity *G*, and the horizontal **pressure at the** crown, may both be transferred to *n* (the intersection of their lines of direction). The resultant is *n A*, which must pass through *A*, and as it represents the direction as well as the intensity of the pressure at *A*, it must be tangent to the curve.

But *A n o* and *A D C* are similar **triangles, and as the** weight is supposed to be uniform, and **as a consequence *A o* = *o c*, it follows** that ***B D*** also equals *B C*, **which is** a well **known** property of the parabola.

The method which we have suggested for finding the curve of equilibrium, is based upon the principle that the horizontal pressure is constant at all points of the arch, and the vertical pressure **upon** any joint is equal **to the** weight **of the portion of the arch** between that **joint and the crown.**

FIG. 77.

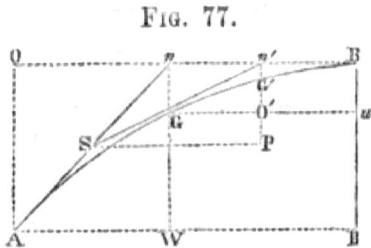

If then this principle be correct when applied to the parabola, it follows that if any joint be taken, as *G*, **and** a line drawn vertically through the centre of gravity of *G **B***, terminated by the line drawn horizontally through the crown; if *n' P* be **made to bear** the same proportion to the weight of *G B*, that *B R* does to the whole **weight** on *A B* or *B O;* then *S P*, which represents the horizontal component of **the pressure** at *G*, should be constant at every part of the **curve, and be** equal to *A W* or ½ *A R*.

To prove that this is the case, and that the parabola conforms to the **rule that we have endeavored** to establish.

Take *B R* **to represent the weight on** *A R* and call it *x*. Also let *A R = y*. Take any point *G*, **and** let *n =* ratio between *G u* and ***A R***. *G u* will therefore be equal to (*n y*).

and $n'P$, which represents the weight on Gu, will be equal to nx. From the equation of the parabola we have $y = \sqrt{px} \therefore ny = n\sqrt{px}$, or since $Gu = ny = \sqrt{p \times Bu}$, we will have $ny = \sqrt{px'}$ (by calling $Bu = x'$ for brevity) $\therefore px' = n^2 px \therefore x' = n^2 x$. But from similar triangles,

$n'O' : O'G :: n'P : PS$, or $n^2 x : \dfrac{ny}{2} :: nx : sp = \dfrac{y}{2} = \dfrac{AR}{2} =$ a constant quantity.

NOTE.—The fact established in the preceding demonstration furnishes a convenient method of describing the parabola by points.

FIG. 78.

Let A and B be two points through which a parabola is to be drawn. Divide AC and BC each into the same number of equal spaces: draw the horizontal and vertical lines through the points of division as represented in figure. Through G (the middle of the first space) draw $Go = Bn$: lay off $om = \frac{1}{2}AC$: draw Gm, and its intersection (s) with the vertical through m will determine a point of the curve, the apex being at B.

Again, on the vertical through m (the middle of $m'C$) lay off $Go' = 2Bn = Bn'$ make $o'm' = \frac{1}{2}AC$ as before, draw Gm' and its intersection with the vertical through m', will determine s': a second point of the curve.

In the same way any required number of points, at equal distances apart, may be determined.

ILLUSTRATIONS

OF

PARTICULAR MODES OF CONSTRUCTION.

As the object in this treatise is to explain the general principles of bridge construction, no attempt will be made in this first part to enter into details, but as the subject would be incomplete without illustrations, outlines will be given of those structures which deserve attention. Some new combinations that might be advantageously employed will be included.

FOOT BRIDGE ACROSS THE RIVER CLYDE.

By PETER NICHOLSON.

FIG. 79.

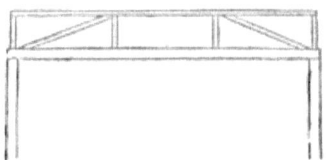

The general arrangement of the supports is represented in the above figure; it consisted of piles, driven into the bed of the stream, across which longitudinal pieces were placed to span the openings; these were strengthened by a framework on top, consisting of two oblique braces with a straining-beam.

The same kind of a frame is much used at present for spanning short intervals; it possesses sufficient vertical strength, but has no counter-bracing, and consequently would be deficient in stiffness. For a foot-bridge, particularly one which does not rest upon stone supports, its flexibility would not be a serious objection. When stone supports are used, every precaution must be taken to prevent vibration, as it breaks the mortar of the joints, loosens the stones, and rapidly ruins the structure.

A bridge built by Palladio across the river Brenta, was precisely similar in principle to the above. This bridge also was built on piles, but the braces and straining-beams, instead of being above the roadway forming part of the balustrade, were placed below and framed into the piles, which extended up to the level of the roadway. This bridge was surmounted by a roof supported by Doric columns, connected below by a light handrail.

BRIDGE OVER THE TORRENT AT CISMORE.

By Palladio. *Span* 108 *feet*.

Fig. 80.

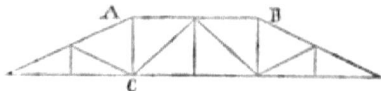

This bridge must have been a good one for small spans. The arrangement is such that the pressures are transmitted to the abutments with very little tendency to produce a change of figure. The rise at the point A, which would be produced by the action of a weight at B, is counteracted by the resistance of the tie AC.

One of the most remarkable designs of Palladio consisted of two parallel or concentric arches connected by diagonal braces.

ILLUSTRATIONS OF PARTICULAR MODES. 143

Fig. 81.

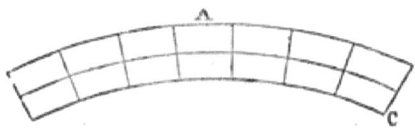

As it appears to have been the first idea of constructing a system of framed voussoirs similar to the arch stones of a bridge, a principle that has been adopted to a considerable extent for iron bridges, and is recommended by Tredgold for **structures in wood**.

In Tredgold's Carpentry, **will be** found a plan for a bridge **of** 400 feet span on this principle.

The objection to this mode of **construction has been** already stated. The only points that bear any considerable portion of the strain are B and D, hence the timber at A **and** C **adds** unnecessarily to the weight. Such an arch **would** undoubtedly be very stiff, and would oppose great resistance to change of figure, but an arch is not the only element required **in the** construction of **a roadway; it is** merely a support from which vertical pieces must **extend** either as posts to support a roadway above, or as ties to sustain one suspended beneath. In either case it is obviously more simple, more economical, more elegant, and more scientific, to make a single arch with the maximum rise, and secure stiffness by counter-bracing between the uprights, a method which has the additional advantage of stiffening the uprights themselves.

Fig. 82.

Another design **from Palladio is** represented above. This **truss** would sustain **a** uniform load, but would not suit for a viaduct without counter-bracing.

BRIDGE ACROSS THE PORTSMOUTH RIVER.

Span 250 feet.

Fig. 83.

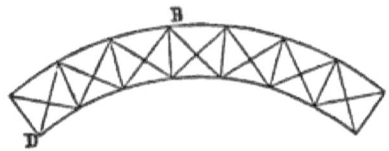

This plan is given on Plate 16 of Tredgold's Carpentry, as a specimen of an American bridge. It is composed of three concentric arcs, connected by radial pieces without either braces or **counter-braces.**

Were the problem given us to arrange a given quantity of **timber in** the most unskilful manner possible, it would be difficult **to select** a plan which would much better fulfil the required conditions. By separating the timbers into **three** arches, and **placing them at** a distance apart, the whole of the strain, or **by far the** greater part, is thrown upon the points A and C, and only one-third of the material is so disposed as to resist it. Again, the **stiffness of** such a system would be little more than one-third **that of a single** arch containing the same material, for **the stiffness being as the square** of the depth in a beam whose **depth is 3, it will be** represented by 9, and in a beam whose **depth is 1, it will be 1.** Hence 3 beams of the depth 1 will only give one-third the stiffness of a single beam whose **depth** was equal to the sum of the **three.**

Colonel Douglass, who gives a description of this work, observes **that the arch is extremely** flexible. This result would necessarily **follow** from **the absence of** counter-bracing.

The quantity of timber **must have** been very great to enable it **to stand at** all, **if heavy** variable loads were drawn over it.

TIMBER BRIDGE OVER THE RIVER DON, AT DYCE IN ABERDEENSHIRE.

Fig. 84.

We were much surprised upon turning to the article Bridge, in the Edinburgh Encyclopedia, to find almost the identical plan of construction which our theory had led us to recommend as best adapted to bridges of large span.

This structure was erected by Mr. James Burn of Haddington, near Aberdeen, **in the** year 1803. The description does **not** inform us in reference to any of the details of construction, and we cannot tell whether the architect wedged **the counter-**braces to increase the stiffness of the truss. The **plan of using** wedged counter-braces appears to have been **but recently in-**troduced, **and** forms a new and important era in bridge construction; even yet, many practical builders do **not seem** to understand their utility.

SCHAFFHAUSEN BRIDGE.

Fig. 85.

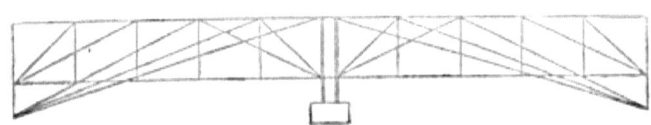

This celebrated structure was built by Ulric Grubenmann, and consisted of two spans, one of 172 feet, the other of 193. It was supported in the interval by a stone pier, which had remained when a former bridge had been swept away. With many excellencies this bridge had also serious defects, and it is certain that a much smaller quantity of timber judiciously arranged would have given far greater strength. Still the principle is an admirable one, and originating as it did with an uneducated village carpenter, certainly displays no ordinary capacity. The supports consist entirely of systems of arch-braces, but the details were too complicated, and the execution evinced considerable timidity. It would seem, from an inspection of the plan, that the design had been conceived of spanning the whole interval at once, as there is a system of arch-braces extending from the abutments towards the centre; but apprehensive that such a long interval would cause the bridge to fail, two other systems of arch-braces were introduced, extending from the extremities towards the centre of each span.

It would have been better, either to have spanned the whole interval by one magnificent truss of 365 feet, which could have been constructed on the arch-brace principle, or else have employed two separate trusses, one for each interval.

A glance at the figure, which exhibits merely the general principle without attempting to represent the complicated details, will show that it was destitute of counter-bracing; and Mr. Cox, a traveller in Switzerland, states that "a man of the slightest weight felt it almost tremble under him, yet wagons heavily laden passed over it without danger."

Upon the principle of the Schaffhausen bridge are the viaducts of the Baltimore and Ohio Railroad, designed we believe by B. H. Latrobe, Esq., Chief Engineer. The arch-brace system is here combined with diagonal ties of iron, by which it is effectually counter-braced. The sizes of the braces are calculated from an exact estimate of the weights they are required to sustain, and the whole arrangement and proportionment evince a thorough acquaintance with the subject, and render the plan admirably adapted to span any interval, or sustain either a uniform or variable load.

LONG'S BRIDGE.

FIG. 86.

The main support of this bridge consists of a system of braces and ties, and in large spans arch-braces are added; keyed counter-braces are also used, and the details are very well arranged.

These bridges have been extensively used on many of the most important railroads of the United States. In the city of Baltimore they are employed at the crossings of most of the streets that intersect the direction of Jones' Falls.

LATTICE BRIDGES.

No plan of bridge construction has met with more general favor amongst engineers and builders than the lattice. Its great simplicity, the ease with which it can be framed, and chiefly its economy, have secured its introduction for viaducts of almost every class. Of late years, however, the frequent failures of these bridges in consequence of heavy transportation, have produced a revolution of sentiment hostile to the plan, and instead of examining into the causes of failure and providing a remedy for the defects which occasioned it, other modes of construction have been adopted at an expense sometimes double that of an efficient lattice structure.

On ordinary roads, and on railways not subjected to very heavy transportation, this plan of superstructure, when well constructed, has been found to possess almost every desideratum. Nevertheless, experience has fully proved that unless strengthened by arch-braces or arches, the capacity of the structure is limited to light loads, and spans of small extent. The public works of Pennsylvania furnish abundant proof of the truth of this assertion; and several railways might be enumerated, on which the lattice bridges have from necessity been strengthened by props from the ground, by arches, or arch-braces added when the insufficiency of the structure was found to require it.

LATTICE BRIDGES. 149

FIG. 87.

The lattice truss in its most simple form consists of two sets of chords, *A B* and *C D*, connected by diagonal ties and braces.

The chords are formed of plank 3 inches × 12 inches, lapped so as to break joint both above and below. The braces and ties are also of 3 inch plank, placed between the chords and pinned with wooden pins at all the intersections. From this description it will be perceived that the truss possesses the merit of simplicity in the highest degree.

The most ordinary carpenter who is able to bore a hole with an augur is capable of constructing it, and the timbers being all of the same size are delivered from the mill in the state in which they are put together; hence no preparatory labor is required, no carefully fitted joints, no bolts, straps, or ties of iron, and consequently it is fair to presume that the lattice principle is the cheapest upon which a truss-bridge can be constructed. If, therefore, its defects can be removed, we see no reason why this mode of construction should not take precedence of all others for ordinary purposes, where economy in first cost is an object of importance.

The elements of the lattice truss consist of horizontal chords, and inclined ties and braces, as represented in the figure.

FIG. 88.

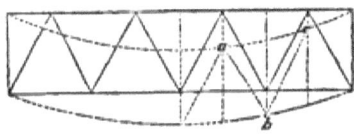

In the case of flexure, the pieces in the direction *a b* suffer compression, and therefore act as braces, and those in the direction *c d* are extended and become ties. The lattice truss therefore possesses this peculiarity, that the ties are all in an inclined position, instead of being perpendicular to the chords, as in other modes of construction.

That this inclined position of the ties is injurious, we are not prepared to prove; although several considerations lead us to suppose that it is less efficient than when the ties are perpendicular. But this point is comparatively unimportant, as it is for very different reasons that we propose a change in the mode of construction.

One of the first defects apparent in old lattice bridges is the warped condition of the side trusses. The cause which produces this effect cannot perhaps be more simply explained, than by comparing them to thin and deep boards, placed edgeways on two supports, and loaded with a heavy weight. So long as a proper lateral support is furnished, the strength may be found sufficient; but when the lateral supports are removed, the board twists and falls. A lattice truss is composed of thin plank, and its construction is in every respect such as to render this illustration appropriate.

A second defect may be found in the short ties and braces at the extremities, which, furnishing but an insecure support, render these points, which require the greatest strength, weaker than any others; this defect is generally removed by extending the truss over the edge of the abutment, a distance about equal to its height, thus providing a remedy at the expense of economy by the introduction of from 15 to 30 feet of additional truss.

Other defects can be mentioned, which are not, however, peculiar to lattice bridges. The ties and braces are of the same size throughout, and consequently no stronger at the point of greatest strain than where the strain is least. The same remark applies also to the chords. Some of these evils can be remedied by slight additions. By bolting arches or arch-braces to the truss, the weak points both of the chords and braces can be effectually relieved. But it would be still better to depend for the power of resisting all the weight upon an arch-brace system, using a light lattice truss only as a counter-brace. This would be a great improvement; but one defect would still **remain**; there would not be sufficient security against warping, although **much more than in the ordinary** method of **construction**.

Fig. 89.

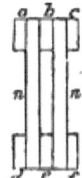

The **double** lattice, as it is called, **consists of three sets of** chords, above **and** below, **as** represented in the **cross-section**, between which two sets of ties and braces are introduced. In comparing this truss with the single lattice, it is evident that it must possess greater power to resist warping, for the timbers $n\,n$ being separated by an interval, will act on the principle of a hollow cylinder, which is much stiffer with a given quantity of material than a solid one. This however is its only advantage, in other respects we think it one of the worst that could be adopted. Whilst the weight of timber from the ties and braces **has been** doubled, the cross-section of the chords has been only increased one-half. A great load of unnecessary timber is placed in the centre, where any weight acts with the greatest leverage, and produces the greatest strain. It is probable that this truss, as usually constructed, possesses less absolute strength with a given quantity of material than any other in common use.

The greatest **improvement that could** be made to this truss, would be to introduce two arch-braces and a straining-**beam**, and the opening between the trusses $n\,n$ would be admirably adapted for the reception of such **a system**.

In lattice bridges a second set of chords is sometimes, perhaps it may be said generally, placed between the first, crossing the second intersections; but as these chords are nearer the neutral axis, they of course act less efficiently than those which are at the top and bottom.

IMPROVED LATTICE.

The following plan of a bridge truss was designed by the author in the year 1840. With even greater simplicity and economy than the ordinary lattice, it appears to be entirely free from its defects; and possessing many of the essential requisites of a good bridge, with a capability of extension to spans of considerable length, it seems to be unusually well adapted to the wants of a community with whom economy is an object.

A well arranged and proportioned structure should possess the following requisites:

1. The cross-section of the chords should be greatest at centre, and least at the ends.

2. The resisting area of the ties and braces should be greatest at the abutments.

3. A system either of counter-braces or of diagonal ties must be introduced, to secure the structure against the effects of variable loads.

4. The timbers of the side trusses should be of such a size, or arranged in such a manner, as to guard against all liability to warp.

5. It is desirable, although not always necessary or practicable, that the pressures should be divided amongst several timbers, so that any defective piece can be readily removed and its place supplied by another, without rendering it necessary to support the bridge during the progress of the repair.

IMPROVED LATTICE

Fig. 90.

In the improved lattice the first two requisites are attained by a system of arch-braces and straining-beams, which is the simplest method of relieving both the chords in the centre, and the braces and ties at the abutments. Arches are preferable, but rather more expensive.

The arrangement of the intermediate timbers is similar to that of the common lattice, and the manner of forming the connections by wooden pins is the same; but the ties instead of being inclined are vertical, a position which is more natural, more efficient, and requires less material.

The braces instead of being single are reduced in size and placed in pairs, one on each side of the tie, which accordingly passes between them, and is pinned at every intersection.

This arrangement secures the third, fourth, and fifth requisites. The inclined pieces, from the manner of their connection, are equally capable of acting as braces or ties, and therefore the truss is counter-braced by a system of diagonal ties, without the necessity of introducing timbers expressly for this purpose, as in most other plans.

The braces being in pairs, with the ties passing between, as in the figure, will possess the stiffness of a hollow cylinder.

Fig. 91.

In this respect it possesses the only good quality of the double lattice, but in a higher degree, for there are here intermediate points a and b, formed by the passage of ties to which the braces are pinned, and which add greatly to the stiffness.

In the ordinary lattice the braces and ties being 3 inches thick, if they were placed upon each other in the same direction, and pinned at short intervals, the stiffness would be nearly in proportion to the square of 6 ; but as they cannot be so arranged, and in fact cross each other nearly at right angles, the flexure of one system is not affected by contact with the other, and the lateral stiffness would only be in the proportion to the square of 3. In the improved plan the braces are made 2 inches by 10, the amount of timber is the same as in the common lattice, but the stiffness would be nearly in the proportion of the square of 7, or five times that of the common lattice.

It is evident, too, that upon the removal of any tie or brace the weight would be sufficiently sustained by the adjacent ones, and repairs could therefore be made without difficulty, an advantage which is not peculiar to this plan, but is possessed also by several others.

In addition to this it may be observed, that the truss does not require to be extended back any considerable distance from the face of the abutment ; there are no short ties as in the common lattice.

The mode of construction which has been designated as the improved lattice, admits of extension to any span to which an arch-braced system is applicable, but is exceeded in the length to which it might be extended by the simple counter-braced arch. In very large spans, whatever be the general arrangement of the timbers of the truss, the whole dependence for the support of the structure and its load should be placed upon the arch-braces and the straining-beams which join the extremities. This system may be connected with a truss on the principle of the improved lattice, by which it will be effectually counter-braced and the parts properly connected ; such an arrangement is represented in the accompanying figure, and could be employed for long spans.

FIG. 92.

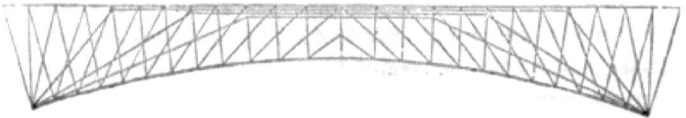

COLUMBIA BRIDGE.

The bridge across the Susquehanna river, at Columbia, consists of a series of spans of about 200 feet each, the whole length of the bridge being about 1¼ miles. This structure consists of a truss composed of braces and ties, strengthened by the addition of an arch, and although the bridge is straight the upper chords are not continued across the piers. From the absence of counter-bracing, it might be inferred that considerable vibration would be produced by the passage of a load.

This is in fact the case; the undulation caused by a passing car can be felt at a distance of several spans. Many of the bridges on the Philadelphia and Columbia Railroad are on the same principle. They are very light structures, but the absence of counter-braces is an objection. The following figure will give an idea of the plan.

Fig. 93.

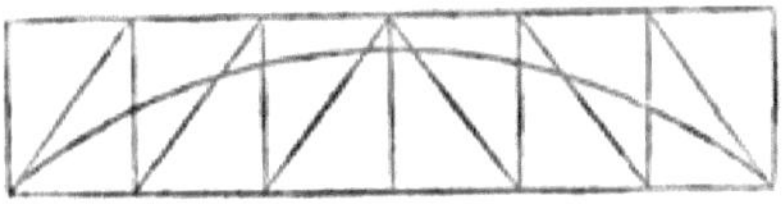

The old bridge across the Susquehanna at Harrisburg, one half of which remains, is similar in principle to that at Columbia, except that it contains heavy counter-braces of nearly the same size as the braces themselves. It is encumbered with unnecessary timber, but in other respects the arrangement is good.

A portion of this bridge was recently carried away by a flood, but it has since been rebuilt. The railroad bridge across the Susquehanna at the same place is on the double lattice plan.

An arrangement something similar in appearance, but differing altogether in principle from the Columbia bridge, and which would possess greater stiffness, consists of a single arch attached to a counter-braced truss. No doubt can be entertained of the ability of the arch to sustain a load if change of figure can be prevented; and the counter-braces would effectually stiffen it, and prevent that injurious vibration to which reference has been made.

Fig. 94.

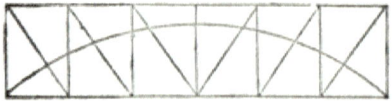

A still lighter truss could be formed by using diagonal ties instead of counter-braces. The vertical pieces would then be in a state of compression, and could be simply notched on the chords without passing through, as is necessary when the strain upon them is one of extension. This arrangement would suit for a bridge when the roadway is on the top chord.

Where the roadway is on the bottom chord, the ties should be iron rods and the counter-braces of wood.

Fig. 95.

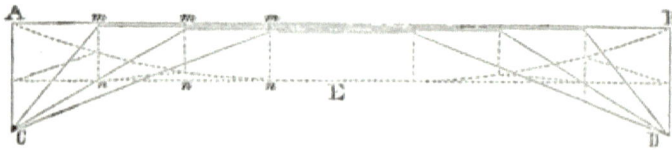

A system of construction applicable to spans of considerable extent, consists of arch-braces counter-braced by a single inverted arch, $A E B$. The arch is attached by iron rods passing through the straining-beams with nuts on the top. The nuts being at the top of the truss would be at all times accessible, the strain could be regulated at pleasure. The long braces at the ends would require intermediate supports.

Instead of the inverted arch, an ordinary counter-braced truss consisting of chords, ties, and counter-braces, without braces, could be used.

Trussed girder bridges which consist of two or more longitudinal timbers, strengthened by iron rods passing beneath them and adjustible by screws, are strong, cheap, and when properly constructed and proportioned, are very efficient. As usually constructed, with two posts dividing the span into three intervals, they are without diagonal rods or braces in the middle interval; this is a defect which should be avoided. For considerable spans, the intervals must be increased in number, and the figure of the truss becomes a polygon, bounded by a straight chord on the upper side, and by one or more iron rods, forming a broken line on the lower side.

An application of this principle, which does not appear to have been made, but which would be useful in many cases, consists in trussing the top instead of the bottom chord. Trussed girder bridges could then be used when the roadway passes through the bridge, as well as when it passes over the top. In this case, the top chord must be well braced laterally, and the ends must be supported by strong posts.

One of the simplest, and, for an iron bridge, one of the cheapest modes of construction, consists in using a single arch, a straight top chord, and vertical posts, or columns connecting the chord and the arch without panel braces or ties of any kind, and without a lower chord. The arch is counter-braced by iron rods extending from the chords over each post to the abutments below the skew-backs, where they are securely anchored into irons passing through the masonry.

Many other arrangements and combinations might be given, but as the object of the author in the first part has been to establish general principles, and not to exhibit details, the reader is permitted to exercise his ingenuity in making other combinations of the elements of bridge trusses, viz., chords, ties, braces, counter-braces, arches, and arch-braces.

PART II.

PREFACE TO SECOND PART.

CONSIDERABLE time has elapsed since the preparation of the former part of this work, during which so many improvements have been introduced into the practice of bridge construction, that a further extension of the descriptions of particular plans seems to be necessary. It is believed that no work has ever been published containing detailed calculations of the strains upon all the timbers which constitute the supporting trusses of a framed bridge, nor has any theory been advanced which furnishes rules by which these strains can be estimated. At the suggestion of several professional friends, who concurred in the opinion that such an addition was a desideratum, the writer was induced to prepare this second part, containing details of most of the arrangements that are exhibited in wooden structures, furnishing illustrations of all, or nearly all, the different modes or forms of calculation that can be required in estimating the strains upon bridges, including combinations of several systems.

There may be, and no doubt will be, differences of opinion in reference to the proper method of estimating the strains where several systems are united, and it is admitted that the solution of the problem must be based upon hypotheses which may not always exhibit the true practical conditions of the case. By supposing all the joints to bear equally, it may be assumed that each system contributes to sustain the load in proportion to its powers of resistance. This never can be practically true; joints cannot be made perfect; and consequently, a calculation made upon this hypothesis must be to some extent erroneous. Notwithstanding this objection there is no better way, and when it is considered that the material is generally elastic, and that those joints which are most tightly compressed at first will yield, and bring others into more intimate contact, the error will, after all, not be of much practical importance, and the results will furnish a safe guide in proportioning the parts of structures.

A proper attention to his professional duties has not allowed the author time for careful revision of the calculations; it is believed, however, that no errors exist that affect any of the general principles, and if mistakes should be found in any of the numerical results, it must be remembered that these calculations are given merely as illustrations, and great accuracy has not been considered necessary.

PREFACE TO SECOND PART.

The construction of the Pennsylvania Railroad requiring a large amount of bridge superstructure, has afforded an opportunity for the introduction of various plans, some of which are new, and all have been so fully tested by experience or based upon such well-tried principles, that they can confidently be relied upon. The description of these plans with some of those in use upon other roads, will furnish a very satisfactory exposition of the present state of the science.

It will be observed that no particular mode of construction is advocated in this work, but efforts are made to illustrate and establish the general principles that must govern the engineer in every case. The following are the most important:

1. In a straight bridge uniformly loaded, and without arches or arch-braces, the strain upon the ties and braces at the middle point of the bridge is almost nothing.

2. The strain at each end upon the same timbers is equal to half the whole weight of the bridge and its load.

3. The strain at intermediate points is proportional to the distance from the middle of the span.

4. Where the bridge is subjected to the action of variable loads, the greatest strain on the ties and braces at the middle of the span is equal to the greatest variable load that can be applied in the interval of one panel.

5. The strain upon the chords is greatest in the middle, and at this point is dependent entirely upon the weight, span, and depth of the truss; the inclination of the

braces has no influence upon the **maximum strain** upon the chords.

6. Where there is only one span, or where in continuous spans the top chords are not connected over the piers, the strain on the end of the chord is nothing, but at the end of the first brace it is equal to one-half the weight on each truss multiplied by the cosine of the inclination of the brace from a horizontal line.

7. Where the weight of the bridge is constant and uniform, counter-braces are unnecessary; but for viaducts, where the load is variable and unequally distributed, they are indispensable.

These principles indicate the most important conditions which every well-proportioned structure should fulfil; any structure built in accordance with them, and having its parts properly proportioned to the strains, must give satisfactory results.

The difficulty of procuring perfectly accurate drawings and details is so great, and the time which the author has been able to devote to this work so small, that he has been compelled in some instances to supply unimportant omissions, by inserting what he believed to be necessary or to have been used in that particular case. For example, the plans forwarded would sometimes be defective in details of lateral bracings, arrangement of flooring, size of plank,&c.; instead of troubling contributors by writing for further explanations, these deficiencies have been supplied as above stated. This course will not be considered objectionable, when it is remembered that the object of this part of the work is chiefly to furnish practical illustrations of the mode of calculation.

For the purpose of comparing the cost of different structures, an estimate has been made in each case for a single track railroad bridge, 16 feet wide between trusses, at the prices now paid for work on the Pennsylvania Railroad; in preparing an estimate from them, an engineer will of course vary these prices to suit his locality.

In making the calculations, only the dead weight has been considered; the effect of the momentum of a passing load depends so much upon contingencies, irregularities of surface, &c., that no attempt has been made to calculate it. It is proper to add from 25 to 50 per cent. in railroad bridges to compensate for these effects.

The deterioration of the resisting powers of timber caused by age must also be considered. After a wooden bridge has been in use for some years, it becomes much weaker than when erected. The allowance made for the safe limit of the resisting power of wood is 1000 pounds per square inch and of iron 10,000 pounds, but it is probable that 800 pounds per square inch for wood, and 8000 pounds per square inch for the tensile resistance of large rods of malleable iron, would be more nearly the true medium between economy and safety; of this, however, every engineer must judge for himself. It is very certain that there is no economy in risk,—an excess of strength is far better than a deficiency.

The tables of data given by Tredgold and others, in which more than 3000 pounds per square inch is given as the strain that timber will bear, is liable to mislead the inexperienced. The writer has observed that students, in making calculations from such data, often arrive at conclusions which a practical man would consider as superlatively ridiculous. Good specimens will bear for a short time even more than this, but great allowances must be made in practice.

PENNSYLVANIA RAILROAD VIADUCT, ACROSS THE SUSQUEHANNA RIVER. (*Plate* 1.)

Description.

THIS structure is located on the line of the Pennsylvania Rail road five and a half miles north of Harrisburg, and crosses the river Susquehanna at an angle of $68\frac{1}{4}°$ with the general direction of the stream. It is supported on 22 piers and 2 abutments. The piers are founded upon cribs, which is the usual method of building in the Susquehanna, and experience has proved it to be perfectly secure, no instance having ever occurred, so far as the information of the writer extends, of the failure of a crib foundation in this river. The cribs of the Pennsylvania Railroad Viaduct consist generally of only one course of timber, 12×12, framed sufficiently wide to extend 2 feet beyond the line of the regular masonry, and sunk sufficiently low to be at least six inches below the surface of extreme low water. The timbers are connected by cross-pieces dovetailed into them. The compartments of the cribs between the cross-pieces are filled with large and small stone, laid compactly without mortar to a level with the upper surface of the crib; upon this is placed the foundation course of the masonry, consisting of very large stones from 20 to 24 inches high, forming a regular course upon which the cement masonry is commenced with an offset of 6 inches.

The character of the masonry is rock range work laid in hydraulic cement. Each pier contains on an average 420 perches of dressed stone; the width of the piers is 6 feet on top and 10 feet at the springing line of the arches, measured perpendicularly to the general direction of the pier. The foundation is protected by rip-rap of loose stone, which becomes continually more solid by the action of freshets, the effect of which is to deposit sand and gravel in the interstices. Each pier is furnished with an ice-breaker, the slope of which is 45°. The ice-breakers are built with the same kind of stone work as the piers themselves; heavy oak timbers are anchored across the face of each, to which the oak facing timbers are securely spiked.

The first pier was covered with bars of cast-iron placed longitudinally, and about 12 inches apart, the spaces between being filled with concrete. An unexpected period of cold weather, immediately after the concrete was laid, caused it to freeze before setting, and a freshet at the same time washed out a portion at the lower end; so that the result was not as satisfactory as under other circumstances it would have been. This mode of facing an ice-breaker is economical, and secure against fire, which might be communicated to the bridge by coals falling upon it, blown off the floor by the force of wind. Ten of the piers are covered with long oak timbers 10 × 10, laid so as to leave openings between of one inch, which, when the timber has become completely dry, will be filled with cement as security against fire. The ice-breakers of the eleven remaining piers are covered with bars of flat iron secured as follows: holes were punched in the iron bars at intervals of 2½ feet, sufficiently large to receive bolts ¾ inch diameter. The bolts were 6 inches long, split for half their length to receive a wedge. Holes were drilled in the stones at proper intervals to receive the bolts. These holes were filled with ordinary mortar of sand and cement, poured in before driving the bolts. This mode appears to answer perfectly; it is much less expensive than lead and more convenient of application.

The foundations of the 22 piers were commenced and carried above water in 16 weeks; but after the eighth pier was founded, operations were suspended for a period of two months not included in this statement. With the exception of a small portion of material that had been previously delivered, the whole of the work on the piers and abutments, including an arcade of three full centre arches of 25 feet span, constituting the eastern approach of the viaduct, was completed in one season.

The whole amount of stone work in the piers and abutments, including three abutments of a single span bridge across the Pennsylvania Canal at the eastern end of the viaduct, and the cribs of the foundations, is 17,000 perches, of which nearly 13,000 perches are built of rock range work. The final estimate for the masonry was $96,355 84.

Each pier cost about $3,500.

Superstructure.

The superstructure of the Pennsylvania Railroad Viaduct is built upon Howe's plan, with the addition of substantial wooden arches. The spans are 160 feet from centre to centre of piers. The following table gives the principal dimensions.

Span from skew-back to skew-back	149 ft.	3 in.
Versed sine of lower arch	20 "	10 "
Under side of skew-back below bottom chord	6 "	
No. of panels, 16.		
Pier panel	3 "	10 "
Distance between wall plates on piers	4 "	8 "
Length of panels	9 "	9 "
Width in clear between arches	13 "	11 "
" " " chords	15 "	5 "
" from out to out of chords	19 "	
" " " " arches	20 "	6 "
Angle of pier and chord	$68\frac{1}{4}°$	
Thickness of pier at right angles to skew-back 10 ft.		

The arches are in 3 segments, the dimensions being at centre 11 + 7 + 11 deep by 9 inches wide, at skew-back 11 + 11 + 11 deep by 9 inches wide.

Hypothenuse of skew-back   33 in
Perpendicular   "   29·2 "
Base   "   15·3 "
Height of truss from out to out of chords   18 ft.

After the 14th span had been raised, a violent tornado occurred, March 27th, 1849, which carried off six spans. These spans were in an unfinished condition. The contractor was engaged at the time in putting in the arches, and as the diagonal braces could not be permanently introduced until after the arches were in place, he had omitted them, except over the piers and in the middle of the spans. The direction of the storm was nearly at right angles to the bridge. The failure commenced at the extreme end which was supported on trestles. The bridge gave way by falling together in the direction of the diagonal. The only arrangement that could have secured the bridge in so violent a tornado, would have been a complete system of diagonal bracing, but the accident occurred before these could be introduced, in consequence of the unfinished condition of the arches.

Experiments were made by the writer to ascertain whether it would have been possible for the wind to carry away the bridge by sliding along the top of the pier or wall plate, but the least friction, in an average of 15 or 20 experiments, was $\frac{7}{12}$ of the pressure, which was sufficient to produce a resistance 4 times as great as the force of the wind upon the exposed surface, at that time, estimating the force of wind at 14 pounds per square foot.*

---

* It is desirable that further experiments should be made to ascertain the force of the wind in **violent storms**. It is probable that it is generally underrated. The writer **addressed** letters to gentlemen who had been engaged in making observations with the anemometer. The most satisfactory answer was given by Professor Bache, who stated that, "on Saturday, August 5th, **1843, at** 8 o'clock P. M., **a tornado passed within a** quarter of a mile of the Observatory (Girard College), and the force of wind was so great as to exceed **the** range of the spring, and to break the wire connecting it with the plate **of the** anemometer; the force required exceeded 42 pounds to the square foot, which was the range of possible movement of the registering arm. The next greatest force of wind was 14 pounds to the square foot, from 4 to 5 o'clock, A.M., **on the 17th Feb.** 1842. From 0 to 5 hours, A.M. on the same

Bills of Materials for one span Pennsylvania Railroad Viaduct.

WHITE OAK.

4	Wall plates	8 × 12	21 ft. long		672
4	Bolsters	8 × 10	17½ "	"	466
12	Braces	7 × 7	19 "	"	932

Total Oak   2,070

WHITE PINE.

84	Chord pieces	5 × 12	39 ft. long		6630
17	" "	5 × 10	39 "	"	2764
8½	" "	10 × 10½	39 "	"	2900
64	Main braces	6 × 7	19 "	"	4262
32	Counter "	6 × 7	19 "	"	2131
64	Lateral "	5 × 6	18½ "	"	2963
44	Floor beams	7 × 14	24 "	"	8624
30	Diagonal braces	5 × 6	23½ "	"	1770
12	Track strings	9 × 11	29½ "	"	2920
72	Arch pieces	9 × 11	27 "	"	16056
60	Purlines	2½ × 4	16 "	"	804
20	"	6 × 12	16 "	"	1920
3200	feet inch boards		10 "	"	3200

Total no. of feet B. M.   59,014
" cubic feet   4,918
No. cubic feet per foot lineal   30·7
Weight of timber per foot lineal, 1105 pounds.

---

day, the mean was 12·6 pounds. From 0 to 11 hours A. M. the mean was 11·4 pounds, and for the day 7·69 pounds."

The above is an extract from the letter of Professor Bache. The observer at the instrument at the time of the tornado was Mr. Lewis L. Haupt (brother of the writer), who says that he has a distinct recollection of the occurrence, and that the wire of the anemometer broke by the force of a sudden blast of wind, at the instant when it registered 30 pounds.

Further information on this subject is very desirable.

Bill of Castings for one span.

30	Bottom chord angle blocks, each 90 lbs.	:=	2,700
30	Top " " " " 86 "	=	2,580
8	Half angle blocks " 58 "	=	464
26	Bottom gibs, 4 holes " $34\frac{1}{2}$ "	=	897
4	" " 2 " " 25 "	=	100
34	Top " 2 " " 20 "	=	680
68	Lateral angle blocks " 13 "	=	884
204	Combination keys " 9 "	=	1,836
68	Lateral bolt washers " 3 "	=	204
236	$\frac{3}{4}$ inch washers " $\frac{7}{8}$ "	=	206
			10,551

Bill of bolts for one span, exclusive of nuts and washers.

236	$\frac{3}{4}$ inch bolts, each	$3\frac{5}{8}$ lbs.	=	855 each	2 ft. 1 in. long.
34	$1\frac{1}{8}$ " " "	$64\frac{3}{4}$ "	= 2001 "	19 " 6 " "
8	$1\frac{1}{4}$ " " "	75 "	= 600 "	19 " 6 " "
20	$1\frac{1}{2}$ " " "	112 "	= 2241 "	18 " 7 " "
16	$1\frac{5}{8}$ " " "	127 "	= 2032 "	18 " 7 " "
24	$1\frac{3}{4}$ " " "	151 "	= 3624 "	18 " 7 " "
				11,353

Arch Suspension bolts.

8	$1\frac{3}{8}$ diam.	each	$32\frac{1}{2}$ lbs.	=	260	6 ft. 6 in.
8	$1\frac{3}{8}$ "	"	50 "	=	400	10 " 1 "
8	$1\frac{3}{8}$ "	"	65 "	=	520	13 " 1 "
8	$1\frac{3}{8}$ "	"	77 "	=	616	15 " 6 "
8	$1\frac{3}{8}$ "	"	85 "	=	680	17 " 1 "
8	$1\frac{3}{8}$ "	"	90 "	=	720	18 "
4	$1\frac{3}{8}$ "	"	93 "	=	372	18 " 6 "

Weight of arch bolts 3568
Total weight of bolts 14,920 pounds.

Weight of nuts for one span.

236	nuts for	¾	inch bolts,	each	½	pound	=	118
68	"	1⅛	"	"	$1\frac{1}{10}$	"	=	75
16	"	1¼	"	"	1·4	"	=	23
40	"	1½	"	"	3·1	"	=	124
104	"	1⅜	"	"	2·0	"	=	208
32	"	1⅝	"	"	4·3	"	=	138
48	"	1¾	"	"	4·6	"	=	221

Total weight of nuts 907 lbs

Estimate of cost of one span.

2070	feet B. M. oak scantling	@	$18 00 per M.	$ 37 26
56944	" " white pine	@	13 00 " "	740 27
10551	lbs. castings	@	02½	263 77
14920	" bolts	@	04	596 80
907	" punched nuts	@	09	81 63
4000	" square feet roofing	@	08	320 00
	Total cost of materials			$2039 73

Workmanship.

160	lineal feet superstructure	@	$5 50	$880 00
236	¾ inch bolts, head at one end	@	10	23 60
34	1⅛ " screws at each end	@	35	11 90
8	1¼ " "	@	45	3 60
52	1⅜ " "	@	55	28 60
20	1½ " "	@	65	13 00
16	1⅝ " "	@	75	12 00
24	1¾ " "	@	85	20 40
4000	square feet metal roofing	@	02	80 00
	painting, &c.			100 00
				$1175 10

Total cost of materials and work, $3214 83. — Cost per foot lineal, $20 00.

Principles of calculation.

Before we proceed to calculate the strains upon the parts which compose **this** truss, **it** is necessary to state distinctly the principles upon which such calculation must be made. It is evident that where two **systems are** connected in the **same** truss, each capable of opposing **a certain** resistance, it will be very difficult **so to proportion the** weight upon each, that the load will be in proportion to the strength of the several portions. **If,** for example, a truss be constructed, and the false works removed before the introduction **of the arches,** if the latter be **bolted to** the posts, the weight **of the whole structure is** sustained **by** the truss itself, and the arches will not bear a single pound, unless they are called into action by an increased degree of settling in the truss. But if the bottom chord of the truss is connected with the arches by means **of** suspension rods with adjusting screws, the whole **truss** may be raised upon the arches, **and** in this case **the latter will bear the whole** weight, and the former none.

Again ; **if** we suppose **the** arches to be connected with the truss before **the removal of** the false works, and the **joints** be equally perfect in both systems, there is a prospect **of a more** nearly **uniform** distribution of the load ; but even **in this** case, we **cannot tell what** portion is sustained by each system, because this will depend upon their relative rigidity. If, for example, one of the systems should experience double the deflection **of** the other, with a given load, the less flexible would **sustain twice as** much as the other when combined, provided they are so nicely adjusted **as to bear** equally when unloaded except with the weight **of the structure.**

In practice, **the** most **convenient way** of securing an equal bearing appears **to** be, to **remove the** false works before the arches are introduced. After the arches are in place, examine the level of **the** roadway, and screw the nuts of the suspension arch rods until **the** truss begins to rise very slightly. As there is necessarily a **certain** degree of elasticity in the truss, it will then be certain that both **systems are** in action.

With all **these precautions, there are** still difficulties in estimating the exact strain upon the parts of a bridge which is sustained by two different systems; for there may be unequal settlement, and the adjustment, however accurately made in the first place, may not long continue. It can, it is true, be tested at any time, by unscrewing the suspension bolts until the truss ceases to settle, and then screwing up again until the truss begins to rise; but it will generally happen that after a bridge has been a long time in operation, the two systems bear very unequal portions, and when the truss itself is not so constructed as to be susceptible of adjustment, the arch almost always sustains the whole weight of the bridge, and its load.

These and many other considerations have led the writer to the conclusion that the best method of constructing bridges is to place the entire dependence upon the arch, using the truss merely as a system of counter-bracing and a support to the roadway.

In the structure now under consideration, either the truss without the arches, or the arches without the truss, would be sufficient to bear the load.

The calculation of the strength will be made on three hypotheses:

1. That the arch sustains the whole weight.
2. That the truss sustains the whole weight.
3. That the arch and truss together form one system.

1st. *Calculation of the strength of the bridge on the supposition that the arch sustains the whole weight.*

The data required in this case are,

Distance of centre of gravity of half truss from abutment	$37\frac{1}{2}$ feet.
Distance from centre of pressure of arch at skew-back, to centre of pressure at crown	$20\frac{5}{6}$ "
Cross section of arches in middle of span	1044 sq. in.
Cross section of arches at skew-back	1188 "
Weight of half-span 1281 pounds per foot,	102,500 pounds
Load on " 2000 "	160,000 "

Total load 262,500 lbs.

$$\frac{262{,}500 \times 37{\cdot}5}{20{\cdot}83 \times 1044} = \text{pressure on arches}$$ per square inch at centre = 453 lbs. or about ⅛ the crushing weight.

The pressure in the direction of the tangent of the circle at the skew-back bears to the pressure in the middle, the proportion of the hypothenuse to the perpendicular of the skew-back; it will therefore be $\frac{453 \times 1044 \times 33}{1188 \times 29{\cdot}2} = 450$ lbs., or almost precisely the same pressure per square inch, as at the middle of the span. It appears, therefore, that the proportion between the cross-section of the arches at the ends and at the centre is very exact, and that the arches alone are sufficiently strong to bear, with perfect safety, 3 or 4 times the greatest load that can ever come upon the bridge.

Strain upon the arch-suspension rods.

There is one suspension rod 1⅜ diameter at each arch, and at each panel, consequently the weight on each portion of the bridge corresponding to the length of a panel; that is, the weight of each lineal portion of 9 feet 9 inches, will be sustained by 4 rods, having a united cross-section of 6 square inches. The greatest weight of the bridge and its load has been found to be 3281 pounds per foot lineal. The weight on one panel will therefore be 31,989 pounds, and the tension per square inch = 5331 pounds, or about one-tenth of the breaking weight.

The arches and suspension rods are, therefore, more than sufficient to sustain the greatest load that can ever come upon the bridge, without any assistance from the truss.

Strain upon the counter-bracing produced by the action of the arch.

This strain will be estimated by supposing one half of the bridge to be loaded with one ton per foot lineal, which is not counterpoised by any weight on the opposite side.

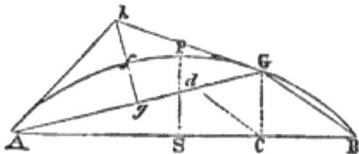

Let the weight on the half span be supposed concentrated at the centre of gravity G.

$B\ C =$ one-fourth span $=$ $37\frac{1}{2}$ feet.
$C\ G = \frac{3}{4}\ P\ S$ (nearly) $=$ 16 "
$A\ G = \sqrt{A\ C^2 + C\ G^2} = \sqrt{112^2 + 16^2} =$ 113 "
$G\ d = \frac{1}{4}\ A\ G =$ $28\frac{1}{4}$ "
$f\ g\ = \frac{13}{16}\ P\ S - \frac{3}{8}\ P\ S = \frac{9}{16}\ P\ S$ (nearly) $=$ 12 "
$g\ h\ = 2\ f\ g =$ 24 "

The weight concentrated at G, is supposed to be 160,000 pounds.

The resultant in the direction $G\ A$, will be $\dfrac{W \times G\ d}{G\ C} = \dfrac{160{,}000 \times 28\frac{1}{4}}{16} = 282{,}500$ pounds.

This force with its equal at A, produces an upward action upon the arch at f, which may be supposed to be resisted by the application of a force at this point.

The required force at f, can be determined from the proportion: $G\ g : g\ f :: 282{,}500 :$ half required force $= \dfrac{282{,}500 \times 24}{56} = 121{,}070$ pounds.

This is the whole amount of force which is exerted upon the arches of both trusses, and which must be resisted by the counter-bracing. The estimate is only an approximation, but it is considerably on the side of safety; for it is impossible that the weight should ever be concentrated at any point G, and the effect of the portion upon $P\ G$ is to assist in keeping down the arch. It is not necessary for practical purposes to make an exact mathematical calculation of this strain: it is sufficient to obtain a near approximation, with the assurance that all errors are on the side of stability. From these considerations, it appears that the upward force upon the arch by the action of the load upon the opposite side, is not more than 121,070 pounds, or 62,000 pounds to each truss.

As there are twelve panels between $A\ G$, there are consequently twelve counter-braces to resist this force, and if each of them sustained an equal portion, there would only be 5170 pounds to each; but a more nearly correct distribution of the pressure is, to allow nothing for the strains at A and G, and double the average for the strain at f; consequently, the greatest possible strain upon any counter-brace, would be less than 11,000 pounds, or only 262 pounds per square inch.

If iron rods had been used for counter-bracing, a cross-section of 1 square inch to each panel would have been an ample allowance.

We will now estimate,

2nd. The strength of the truss itself without the arch.

DATA.

The distance of the centre of gravity from the point of support, is 39 feet.

The distance from middle of upper to middle of lower chord, is 17 "

The resisting cross-section of upper chord, is 205 sq. in.

The combination-keys of the lower chord cut off one-half inch on each side of each chord plank, or three inches. The splice at each panel $4\frac{1}{2}$ inches; the combination-bolt about equivalent to $1\frac{1}{4}$ inches; there remains, therefore, for the actual resisting area of the lower chords 135 "

We will assume that timber should never be subjected to the action of a weight that could be sufficient to impair the elasticity; and that within the elastic limits, the resistances to compression and extension are equal. We will also leave out of view the additional strength which is derived from the continuity of the spans; since this advantage would not be possessed by the spans at the extremities, or by an isolated span of the same extent. The calculation, therefore, will be made, as such calculations always should be made, under the most unfavorable circumstances.

As each upper chord presents a resisting area of 205 square inches, and each lower chord an area of 135 square inches, and as the resistances per square inch are supposed to be equal, the position of the neutral axis will not be in the middle, but must be determined by the condition that the moments of the resistances, or the products obtained by multiplying each area by the distance from the neutral axis, shall be equal.

If x represent the distance of the neutral axis from the middle of the lower chord, $17-x$ will be its distance from the middle of the upper chord.

Let R represent the average strain per square inch upon the lower chord; as the strains are in proportion to the distance from the neutral axis, the strain per square inch on the upper chord will be expressed by $R \dfrac{17-x}{x}$.

The equation of equilibrium will be

$$R \frac{17-x}{x} \times 205 \times (17-x) = 135\ R\ x$$

$$70\ x^2 - 6970 = 59245$$

$$x = 9.4$$

$$\frac{17-x}{x} = .808.$$

The equation of moments will be

$$7.6 \times .808\ R \times 205 + R \times 135 \times 9.4 = 39 \times \frac{262.500}{2}$$

$$2527\ R = 511875.0$$

$R = 2000$ pounds per square inch (nearly).

It appears, therefore, that the strain upon the bottom chords, on the supposition that the arches are omitted, and the bridge loaded with a train of locomotives producing a weight of one ton per lineal foot, would be 2000 pounds per square inch.

Strain upon the ties.

It has been shown that the greatest strain upon the ties at the middle of a bridge is equal to the greatest load that can ever come upon one **panel**; consequently, it will be the same as was determined for the weight upon the arch suspension rods, or 31,989 pounds.

There are 4 rods at each panel, 2 to each truss.

The rods at the middle of the bridge are $1\frac{1}{2}$ inches diameter. The united **cross-section** of the 4 rods will be **7** square **inches**, and the strain per square inch 4,569 pounds.

The end rods sustain the weight of one-half the bridge and its load. Continuing the same hypothesis as formerly, the weight of the half-span and its half-load has been found to be 262,500 pounds.

This is sustained by 4 rods, each $1\frac{3}{4}$ diameter, the united cross-section of which will be 9·6 square inches. And the strain therefore 27,344 lbs., or one-half the breaking weight.

The pressure upon the braces will **bear to the** strain upon **the rods, the** proportion of **the** diagonal **of the panels to** the **perpendicular;** this proportion is, in the **present case, as** 19 : 16. We have therefore

For **the pressure on the middle braces** $\dfrac{31{,}989 \times 19}{16} =$

38,000 pounds **nearly**.

The cross-section of **the braces in the** middle is **168** square inches.

The pressure per square inch on the middle brace is 226 **pounds.**

For the pressure upon the end **braces,** we have

$$\frac{262{,}500 \times 19}{16} = 311{,}718 \text{ pounds.}$$

The 2 **trusses** at the ends contain **6** braces, **4 of which are** of pine, 6 × 7; **the** others are of **oak, 7** × 7.

The united cross-section will therefore be 266 square inches. The pressure per square inch **will** be 1172 pounds.

It is necessary **to** inquire **whether** this pressure of 1172 **pounds** per **square** inch will **cause** the brace **to** yield by flexure.

The braces at the ends are **three in number,** placed side by **side, and supported in the** middle **by the** counter-brace. Two cases present themselves for consideration.

1*st.* Flexure may take place in the direction of the plane of the truss; in which case the resistance will be due to 2 braces 6 inches deep, 7 inches broad, and $9\frac{1}{2}$ feet long; 1 brace 7 inches deep, 7 inches broad, and $9\frac{1}{2}$ feet long.

2*nd.* Flexure may take place in a direction perpendicular to the plane of the truss, in which case the resistance will be due to 2 braces 7 inches deep, 6 inches wide, and 19 feet long; 1 brace 7 inches deep, 6 inches wide, and 19 feet long.

The formula which expresses the extreme limit of the resistance to flexure when the material is white pine, is

$$W = \frac{9000\ BD^3}{l^2}$$ in which l is in feet, and the other dimensions in inches.

By substituting the proper dimensions, we have in the first case

For the two 6 × 7 pieces, $W = \dfrac{9000 \times 7 \times 6^3 \times 2}{9\cdot 52} = 301{,}561$

For the one 7 × 7 piece, $W = \dfrac{9000 \times 7 \times 7^3}{9\cdot 52} = 239{,}434$

Total, representing the extreme limit of resistance, 540,995

The actual strain upon the braces at the end of one truss is 155,876 pounds, which is sufficiently far below the limit to insure perfect security against flexure in the first case.

For the second case, in which flexure is supposed to take place in a direction perpendicular to the plane of the truss, we have

For the two 6 × 7 pieces, $W = \dfrac{9000 \times 6 \times 7^3 \times 2}{19^2} = 102{,}600$

For the one 7 × 7 piece, $W = \dfrac{9000 \times 7 \times 7^3}{19^2} = 59{,}850$

Total, representing the extreme limit of resistance, 162,450

This calculation, it must be remembered, is based on the supposition that the truss is not assisted by arches, and that its load, independently of the weight of the structure, consists of a train of locomotives of the largest class, extending entirely across the span. These are hypotheses which will never express the actual condition of the Susquehanna bridge; but it is proper to examine them, as bridges are frequently built on similar plans without arches, and on roads over which very heavy trains are carried.

The result of the calculation proves that, in the first case, where the counter-brace forms an intermediate support, and reduces the length of the unsupported portions of the **braces to** 9½ feet, flexure cannot take place in the plane of the truss.

In the second place, where there is no intermediate support, and the three braces are supposed to yield laterally in a direction perpendicular to the plane of the truss, the resistance is not sufficient, and flexure might take place under the hypotheses assumed, unless, by the addition of keys and bolts, the three braces are made to act as one piece, in which case the formula will give

$$W = \frac{9000 \times 6 \times 21^3}{19^2} = 1,386,000 \text{ nearly,}$$

which is more than double the stiffness in the first case, and 9 times as great as the maximum strain.

The conclusion, therefore, is, **that in a truss** constructed upon these principles, **but without** arches, **it is highly important** that the braces at the ends of the spans should be stiffened laterally **by** bolts and keys.

Strain upon the floor beams.

The floor beams are 7×14, **placed 3 feet** 8 inches from centre to centre; the interval between chords is 15½ feet; the **greatest weight** upon any floor beam would be equivalent to 4½ tons applied at the centre.

On the supposition that the deflection is $\frac{1}{40}$ inch to 1 foot,

$$W = \frac{B D^3}{\cdot 0125\, l^2} = \frac{7 \times 14^3}{\cdot 0125 \times (15 \cdot 5)^2} = 6402.$$

The actual weight is 9000.

The deflection produced by this weight would be

$6402 : 9000 :: \frac{1}{40} : \frac{1 \cdot 4}{40}$ or $\frac{14}{400}$ of an inch per foot in length.

The actual deflection caused by the passage of a locomotive, allowing 6 tons weight upon a pair of driving wheels, will be $\frac{14}{400} \times 15\cdot 5 = \frac{217}{400}$, or one half inch nearly.

To determine the strain upon the fibres, we must use the formula $R = \frac{3\,w\,l}{2\,b\,d^2}$ in which all the dimensions are in inches, and R expresses the maximum strain per square inch,

$$R = \frac{3 \times 9000 \times 15\cdot 5 \times 12}{2 \times 7 \times 14^3} = 1830 \text{ pounds per square inch.}$$

The effect of the track strings in distributing the pressure over several adjacent floor beams, has not been taken into consideration, because it is not safe to make any allowance in ordinary cases. As a locomotive of the first class occupies considerable space in the direction of the track, several beams may be loaded with equal weight at the same time, **and** one could **not** assist another; besides, there must be joints in the track strings, and at the joints the beam must bear the whole strain.

This weakness is compensated in the Susquehanna Bridge by making the floor beams at the joints two inches wider than at the intermediate points; the strain upon them will therefore be less than that previously determined, **in the proportion of 7 to 9**; it will consequently be 1424 pounds per square inch.

Strain upon the counter-braces.

It has been shown, that when the load upon a bridge is uniform, the counter-braces do not act. Their office is to resist the upward action, produced by an unequal distribution of the weight. The greatest variable load is estimated at one ton per lineal foot, or 160,000 pounds to the half span; the effect of which is resisted by the counter-braces of the opposite side. The conclusion arrived at in considering the subject of the action of the counter-braces, was, that the greatest resistance which it was ever necessary for any one of them to oppose, was equal to the pressure upon the braces of the middle panel caused by the action of the greatest variable load. The **greatest** load upon one **panel is** ten tons, or 20,000 pounds;

the strain in the direction of the brace is $\dfrac{20{,}000 \times 18}{10} =$ 36,000. The cross-section of the 2 braces is 84 square inches. The pressure, 430 pounds per square inch.

Lateral braces.

The greatest strain upon the lateral bracing of a bridge, would be that caused by the action of the wind in a violent tornado. It is probable that this force is far greater than it is usually estimated. The observations of the writer at the Susquehanna Bridge, during the tornado which caused the loss of six of the unfinished spans, led him to believe that the direct effect of the storm was increased by reflection from the surface of the water. It appears reasonable to suppose that if the direction of the wind is such as to strike the surface of the water at an angle of reflection, it must be thrown upwards, and its effect would be to augment the pressure upon any surface exposed to its action. In covered bridges particularly, it is probable that this reflected current acting against the underside, and in opposition to gravity, might so reduce the weight of the structure as to cause it to be blown off the piers. The possibility of this contingency we propose to examine after the direct effects have been considered. In the absence of positive information, the necessary data will be assumed.

The Pennsylvania Railroad Viaduct is designed to be left entirely open. The amount of side surface in the chords and braces is 688 square feet to each, and as the wind can act upon both trusses, the surface presented will be 1376 square feet.

There are two sets of lateral braces, one at the lower, the other at the upper chord. As the upper set is subjected to a greater strain than the lower, the calculation will be limited to it. The upper set of lateral braces may be considered as resisting the force of the wind on one-half the side surface of the truss (688 square feet), and the force of the roadway and railing ($6 \times 160 = 960$ square feet). Allow for obliquity of direction, which would increase the surface 352, making a total of 2000 square feet.

If we suppose that a storm could be so violent as to cause a pressure of 30 pounds per square foot, the whole lateral force would be 60,000 pounds. The force upon the lateral brace rods would be, at each end, 30,000 pounds; and as these rods have a cross-section of $1\frac{1}{4}$ square inches, the strain per square inch would be 24,000 pounds, or about one-half the breaking strain.

This estimate is probably beyond the extreme limit of perfect safety; for the force of the wind has been rated twice as great as is given in tables of violent storms. No allowance has been made for the assistance of the flooring boards, which is very considerable, and the diagonal braces, which transfer a considerable portion of the pressure to the lower chords and the lower diagonal braces.

The strain upon the lateral braces will be to the strain upon the rods, in the proportion of the diagonal of the panel to the perpendicular, or as $18 : 15\frac{1}{2}$; it will therefore be 36,000 pounds, or 1200 pounds per square inch.

The lateral braces 5 × 6, 18 feet long. The limit of resistance to flexure, as determined by the formula $\dfrac{9000\,b\,d^3}{l^2}$ will be 29,988 pounds. The strain under the present hypothesis, is 36,000 pounds; consequently, the lateral braces would require support in the middle.

If supported at the middle point, the length of the unsupported portions would be reduced to 9 feet, and the limit of resistance would be 119,952 pounds.

In which case there would be a considerable surplus of strength.

It appears, therefore, that for security in a very violent tornado, the lateral braces should be supported in the middle of their length, which is the case in the Susquehanna Bridge.

Strain upon the diagonal braces.

If the bridge is in a perpendicular position, the strain upon the diagonal braces will result from the force of winds upon the side trusses. In the Susquehanna Bridge, as the sides are open, and there is a close parapet 6 feet high on the top of the bridge, the centre of gravity of the surface will be very high, and may be taken at the level of the top chord.

Estimating the side surface at 2000 square feet, and the force of a storm at 30 pounds per square foot, the total force will be 60,000 pounds horizontally.

The strain in the direction of the diagonal will be to this horizontal strain in the proportion of 23 to 15, it will consequently be 92,000 pounds.

The diagonal braces are 5 × 6, and 23 feet long.

If they are bolted together at their intersections, the resistance in the direction of a plane passing through them will be considerably greater than in a perpendicular. The estimate should of course be made in the direction of least resistance.

The least lateral resistance, when the two braces are bolted together, will be twice that of a post 23 feet long, 6 inches broad, and 5 inches deep. The limit of resistance, as determined by the formula $W = \dfrac{9000\, b\, d^3}{l^2}$, will be

$$W = \frac{9000 \times 6 \times 5^3}{23^2} = 12,500 \text{ pounds}$$

nearly; as this is the force that will actually cause the brace to yield by flexure, or the extreme limit of resistance, it will not be safe to allow more than 8000 pounds to each brace, if it acts singly, or 16,000 pounds if each pair is bolted; and as the force to be resisted is 92,000 pounds, there will be required in the first case 12 pairs of diagonal braces, and in the second case 6 pairs.

It appears, therefore, that a large amount of diagonal bracing is necessary to resist a strain capable of producing a pressure of 30 pounds per square foot. These braces cannot be permanently introduced until after the arches are in place, and the loss of the six spans at the Susquehanna Bridge was in consequence of the unfinished condition of the bridge, which did not admit of the permanent introduction of a sufficient number to resist the effects of the sudden and violent tornado to which it was exposed.

Resistance to sliding upon the supports.

A bridge which is securely braced diagonally, laterally and vertically, may yield to the force of the wind by sliding upon the tops of the piers, or wall-plates. The possibility of failure from this cause, in the case of the Susquehanna Bridge, we will proceed to examine. Assume that the force of the wind is 30 pounds per square foot, and that an additional force is exerted by reflection, equal to 10 pounds per square foot, acting vertically in opposition to gravity, and reducing to this extent the weight of the bridge. These conditions will probably require a greater power of resistance than will ever be actually necessary in service.

The experiments of the writer on the friction of wood upon wood, and of wood upon stone, give $\frac{7}{12}$ of the pressure, as the least resistance to sliding.

The width of the floor of the bridge being 24 feet, and the span 160 feet, the number of superficial feet of surface will be 3840, and the upward force of the wind, at 10 pounds per square foot, will be 38,400 pounds.

The weight of one span without a load is 205,000 pounds; deducting the first result, the difference, which is the resisting weight of the span, will be 167,400 pounds; $\frac{7}{12}$ of this weight will give for the friction, or resistance to sliding, 97,650 pounds.

Estimating the maximum amount of surface in the open bridge at 2500 square feet, and the force of wind at 30 pounds, the pressure would be 75,000 pounds, or 22,650 pounds less than the resistance.

It is proper to conclude, therefore, that the bridge, if not weather-boarded on the sides, can never be blown over by a tornado, provided the lateral and diagonal braces are sufficient.

Anchor-bolts are used to fasten the bridge to the piers, and in addition to this, the masonry has been extended as high as the top chord. Without these precautions the bridge appears to be secure; with them no storm can possibly have power to move it.

Power of resistance of the Susquehanna Bridge, on the supposition that the arches and truss form but a single system.

The strength of each system has now been separately examined, and calculations made of the strains upon every part. It remains to consider their action when united, on the supposition that each contributes to sustain the load in proportion to its powers of resistance. This hypothesis does not express the exact conditions of the problem, but it is the only one that can be assumed, and the deviation from the truth is much less than some engineers suppose. The objection which is urged against a combination of two systems, is, that either one or the other must measurably sustain the whole weight, and the one which is not active, is merely an incumbrance to the other. This objection is plausible, but the assumption on which it is based is contrary to truth. One system, as has already been stated in the general consideration of this subject, could bear the whole weight only on the supposition that it was absolutely incompressible, which is not the case. It is very clear, that if either system is overloaded, it will yield, and the other will be brought into action.

A very brief consideration of this case will be sufficient.

The strain upon the braces at the middle of the span, will be the same as in the former case. As the roadway is on top, and the weight of a passing load is transmitted to the lower chords and arches through the medium of the braces, the latter must be capable of sustaining the weight upon one panel.

The strain in this case has already been found to be

For the middle braces, 226 pounds per square inch.

The force transmitted by the braces to the lower chords is not resisted by the ties only, but also by the arch-suspension rods, the united cross-section of which is 13 square inches.

The weight to be sustained is 31,989 pounds.

The strain per square inch is 2,451 "

The vertical force at the ends is resisted by the arch and the braces. If the hypothenuse of the skew-back expresses the pressure in the direction of the arch, the perpendicular will represent the horizontal, and the base the vertical pressure, as also the proportions of the resisting surfaces. As the hypothenuse in the present case is 33, and the base 15, the resisting surface which the arches are capable of opposing to the weight, will be $15 \times 9 \times 4 = 540$ square inches.

The cross-section of the braces is 266 square inches, which expresses the resisting surface in the direction of the diagonal, and $\dfrac{266 \times 16}{19} =$ portion, which opposes a vertical resistance $= 224$ square inches.

The whole area which resists the pressure at the ends of the bridge, may therefore be estimated at 806 square inches, and as the weight has been found to be 262,500 pounds, the pressure per square inch at the ends will be 326 pounds.

The portion of this weight sustained by the truss will be $266 \times 326 = 86,716$ pounds, to resist which the cross-section of the 8 rods is 16·6 square inches, and the strain per square inch 5,224 pounds.

Estimate of the longitudinal strains.

In estimating the resistance of a truss composed of two systems, it is not correct to assume that it will be equal to the sum of resistances, which each separately would be capable of opposing, because the strains upon the several portions will depend upon their distances from a common neutral axis.

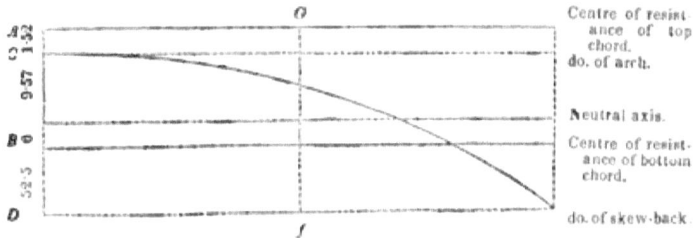

If, for example, we consider the truss $A B$ without the arch, the upper and lower chords sustain nearly equal strains, and the neutral axis will be nearly equidistant from both; but if an arch be added, having a centre of resistance at a considerable distance below the line B, the neutral axis will be brought lower, and the **proportions may be such** that it will fall exactly upon the lower **chord, in which case** the latter will sustain no portion whatever of the strain; **as a** consequence, **the resistance of the system will not be** equal to the sum of the resistances **of its** component parts.

The **first** step in the calculation must therefore be, to determine **the areas of the resisting** surfaces, and the position of the **neutral axis.**

DATA.

The height of truss, from middle of top to middle of bottom chord, is 17 feet.

From middle of bottom chord to middle of skew-back, $5\frac{1}{4}$ feet.

From middle of top chord to middle of arch at crown, $1\frac{1}{4}$ feet.

Resisting area of top chord, A, 410 square inches.

Resisting area of bottom chord, B, 270 square **inches.**

Resisting area of arch, at crown, C, 1044 **square inches.**

Resisting area of perpendicular of skew-back, D, **1051** square inches.

Let $x =$ distance of neutral axis from $C, = x$.

Dist. of A from neutral axis, $x + 1\cdot 25$.

" B " $17 - (x + 1\cdot 25) = 15\cdot 75 - x$.

" D " $15\cdot 75 - x + 5\cdot 25 = 21 - x$.

The pressures upon **the** resisting surfaces will be in proportion to **their** distances **from** the neutral axis, and if R represents the greatest strain per square inch upon the resisting surface, which is at the greatest distance from the axis, which **in** the present **case is** D, the pressure **upon** the other surfaces **will** be per square inch.

Upon the resisting surface A $R \times \dfrac{x + 1\cdot 25}{21 - x}$

"   " B $R \times \dfrac{15\cdot 75 - x}{21 - x}$

"   " C $R \times \dfrac{x}{21 - x}$

The resistance of each surface will be

$$A = AR\left(\frac{x+1\cdot 25}{21-x}\right)$$

$$B = AR\left(\frac{15\cdot 75-x}{21-x}\right)$$

$$C = AR\left(\frac{x}{21-x}\right)$$

$$D = AR$$

The equation of moments will be

$$AR\left(\frac{x+1\cdot 25}{21-x}\right)(x+1\cdot 25) + CR\left(\frac{x}{21}-x\right)x = BR$$

$$\left(\frac{15\cdot 75}{21-x}\right)(15\cdot 75-x) + DR(21-x),$$

Whence $410(x^2 + 2\cdot 5x + 1\cdot 56) + 1044 x^2 = 270(248 - 31\cdot 5 x + x^2) + 1051(441 - 42x + x).$

Reducing $133 x^2 + 53,672 x = 529,811$

$x = 9\cdot 75$ feet, very nearly.

Consequently the distance of the neutral axis from the middle of the upper chord, will be 11 feet.
and from the middle of the lower, 06 "

We are now prepared to determine the strain upon each of the resisting surfaces.

Continuing the former notation, in which R represents the average strain per square inch on the surface D, the strains upon the other surfaces will be, per square inch,

$$\text{At } A = R \times \frac{11}{11\cdot 25} = \cdot 977 R$$

$$\text{At } B = R\frac{6}{11\cdot 25} = \cdot 533 R$$

$$\text{At } C = R\frac{9.75}{11\cdot 25} = \cdot 866 R$$

$$\text{At } D = \underline{\qquad}$$

The whole strain upon each surface in terms of R, will be

Upon							
	A	=	410	×	·977 R	=	400 R
"	B	=	270	×	·533 R	=	144 R
"	C	=	1044	×	·866 R	=	904 R
"	D	=	1051	×	1·000 R	=	1051 R

The weight upon the half span has been shown to be 262,500 pounds.

The distance of the centre of gravity from the skew-back is 38 feet.

The moment of the weight will therefore be $262{,}500 \times 38 = 9{,}975{,}000$.

This is resisted by

$$\left.\begin{array}{rcrcrl} 400 & R & \times & 11 & = & 4{,}400\ R \\ +\ 144 & R & \times & 6 & = & 864\ R \\ +\ 904 & R & \times & 9{\cdot}75 & = & 8{,}815\ R \\ +\ 1{,}051 & R & \times & 11{\cdot}25 & = & 11{,}824\ R \end{array}\right\} = 15{,}909\ R.$$

The equation of equilibrium between the acting and resisting forces will then be

$$15{,}909\ R = 9{,}975{,}000$$
$$R = 627 \text{ pounds.}$$

The average strain per square inch, upon each of the resisting surfaces, will therefore be,

Upon the upper chord $627 \times {\cdot}977 = 613$ pounds.
"   lower   " $627 \times {\cdot}533 = 334$ "
"   arch at crown, $627 \times {\cdot}866 = 543$ "
"   " skew-back, $627 \times 1 = 627$ "

The calculations for the Susquehanna Viaduct have been extended to minute details, in order to illustrate the principles of calculation, and test the strength of every part of this important structure.

The following summary of results, will be convenient for reference.

GENERAL SUMMARY.

No. of feet B. M. white pine, in one span,   56,944 feet.
"   "   "   oak,   "   2,070 "
"   cubic feet timber, per lineal foot,   30·7 "
Weight of timber per lineal foot, in pounds,   1,105 pounds.
"   cast-iron, in one span,   10,551 "
"   "   per lineal foot,   66 "
"   bolts, exclusive of arch-bolts and nuts,   11 352 "

RAILROAD VIADUCT.

Weight of bolts, exclusive of arch-bolts, per
    lineal foot,   71 pounds.
   "   arch bolts,   3,568   "
   "   nuts,   907   "
   "   finished bridge, **per lineal foot,**   1,281   "
   "   loaded bridge,   "   3,281   "
Cost of material for one span,   . $2,039 73
   "   workmanship, for one span,   1,163 10
Total cost for one span,   ——— $3,202 83
Cost per foot lineal, without roof,   20 00
If the arch sustains the whole weight, the pressure at the
    crown will be   453 lbs. sq. in.
If the arch sustains the whole weight,
    the pressure at the skew-back will be   450   "
The strain on suspension rods,   5,331   "
      "   counter-brace,   262   "
Resisting area of upper chords,   410 sq. in.
   "   "   lower chords,   270   "
If the arch is omitted, the strain on the
    lower chords loaded, will be   2,000 lbs. sq. in.
Strain upon the ties, in the middle,   4,569   "
   "   "   "   at the ends,   27,344   "
   "   "   braces, in the middle,   226   "
   "   "   "   at the ends,   1,172   "
   "   "   counter-braces,   430   "
   "   "   floor beams,   1,830   "
Greatest possible strain upon lateral braces, 1,200   "
   "   "   "   "   rods,   24,000   "
   "   "   "   "   diagonal "   133   "
No. of pairs of diagonals required for one span, 6.
Power of wind on side surface,   75,000 lbs.
Resistance to sliding,   97,650 "

Strains upon the parts when both systems are united.

Upon the braces, in the middle of the span, 226 lbs. sq. in.
   "   suspension rods,   2,451   "
   "   braces at the ends,   326   "

Upon the suspension rods at ends,   5224 lbs. sq. in
" upper chord,   613 "
" lower chord,   334 "
" arch in centre,   543 "
" " at skew-back,   627 "

COVE RUN VIADUCT. (*Plate* 3.)

This design was prepared for a bridge across Cove Run, on the Pennsylvania Railroad, but in consequence of peculiarities of location, another plan was submitted; it is inserted here in consequence of its simplicity.

Description.

The span is 50 feet from skew-back to skew-back.
Width from out to out, 9 feet.
Height of truss from out to out, 10 feet.
Number of trusses, 2.
The upper chord is a single timber, 12 × 12, of white pine.
The posts are of locust, 6 × 6, supporting the upper chord.
The arches are composed of rolled plates; each arch consists of 8 plates, 2 × ¾, with a space in the middle of two inches, the upper and lower portions being separated by blocks of cast-iron. The lower chord is of rolled iron, and is designed not to resist the thrust of the arch, but to connect the system of counter-bracing.

The lateral braces are of wood, supported by angle-blocks of cast-iron, and connected by rods ⅞ inch in diameter. The counter-brace rods are one inch diameter, passing through angle-blocks on the upper chord, and connected with the lower chord by means of eyes passing around the lower lateral brace rods.

COVE RUN VIADUCT.

Bill of Materials for one Span.

Cast-iron.	Cubic inches.
40 blocks between arches, each 7½ cubic inches	300
36 post plates, each 30 cubic inches	1,080
40 coupling plates for arches, each 20 cubic inches	800
20 angle-blocks, **for** counter-brace rods, each 14 cubic inches	280
40 angle-blocks, **for** lateral brace **rods, each** 46 cubic inches	1,840
40 **washers for** lateral bolts, 3 cubic inches	120
6 **angle-blocks,** for diagonal brace rods, each 14 cubic inches	84
6 washers for diagonal brace rods, each 3 cubic inches	18
4 skew-backs, **each** 300 cubic inches	1,200
Total **cubic inches of** cast-iron	5,722

Weight in pounds, 1431.

Bill of Malleable Iron.

32 **plates,** or 1728 lineal feet of rolled iron, 2 × ¾ for arches	31,104
20 counter-brace rods ⅞ diameter, 12 feet long, 88 cubic inches	1,760
10 diagonal tie rods 1 inch diameter, 14 feet long, 132 cubic inches	1,320
20 lateral brace rods ⅞ inch diameter, 8⅓ feet long, 61 cubic inches	1,220
18 **bolts** through posts 1¼ inch diameter, 13 inches long, 16 cubic inches	288
4 skew-back bolts 1½ inch diameter, 16 inches long, 28 **cubic** inches	112
80 **short bolts for** arches, 12 inch by ¾ diameter, 5¼ cubic inches	420
4 tie plates 3 × ½, 50 feet long, **900 cubic inches**	3,600
152 nuts, each 3 **cubic** inches	456
44 nuts, each 9 cubic inches	396
Total cubic inches	40,676

Weight in pounds, 10,169.

Wood.

2 top chords	12 × 12, 54 feet long,	B. M. =	1,296
36 lateral braces	4 × 5, 8½ "	" =	510
44 oak cross-ties	8 × 8, 10 "	" =	2,346
22 locust posts	6 × 7, 10 "	" =	770
			4,922

Weight, 14,766.

Recapitulation.

Weight of cast-iron in bridge,   (single span,)   1,431 lbs.
  "   malleable iron in bridge,   "   10,169 "
  "   timber including cross-ties,   14,766 "

Total weight, 26,366

Weight per foot lineal of bridge 528 lbs., or ¼ ton nearly.

Calculation.

The action of the truss consists in counter-bracing the arch, it has of itself no sustaining power. The abutments resist the thrust.

Allowing, as usual, **one** ton per foot lineal as the maximum load on a **single track** railroad bridge, the weight upon the **half-arch will be 63,183** pounds.

The height from middle of skew-back, to middle of arch at curve, **is 10 feet.**

The distance **of the centre of gravity** from the abutment, **12 feet.**

The horizontal strain at the **centre** will be

$$H = \frac{63,183 \times 12}{10} = 75,819 \text{ pounds}$$

The strain at the skew-back will be

$$\frac{75,819 \times \sqrt{12^2 + 10^2}}{12} = 101,000 \text{ pounds nearly, or one-third more}$$

than at the middle of the span.

To resist this we have the four arches, the united section of which will be 24 square inches, or 4200 lbs. per square inch, being nearly one-fourteenth of the crushing weight.

Estimate of cost of Cove Run Viaduct.

1,431 lbs. castings @ 2½ cts.	$ 35 77
10,169 " malleable iron @ 3½ cts.	356 91
1,800 feet pine scantling @ $15 per M.	
(board measure)	27 00
2,400 " oak scantling @ $20 per M. do.	48 00
770 " locust " @ $40 "	30 80
Making 152 bolts, at an average of 15 cts.	22 80
Workmanship, 52 feet @ $6 per foot	312 00
Total cost	$833 28

Total cost per foot lineal, $16 00.

IRON BRIDGE ACROSS HARFORD RUN, BALTIMORE.
(Plate 5.)

This bridge was built by **Daniel** Stone, of Massachusetts. It furnishes a good illustration of the application of the principle of Howe's truss to an iron bridge. There is much simplicity in the arrangement of the details.

Description.

This bridge is 32 **feet long and** 66 feet wide. It is supported by 5 trusses, at intervals of 16½ feet from centre to centre. The two outside trusses are above the roadway, serving both as a support and as a railing; the 3 intermediate trusses are entirely below the roadway. In the middle of the bridge is a railway track, which is directly over the middle truss, one rail being on each side, and the pressure is distributed over the trusses by means of two longitudinal timbers immediately below the transverse floor beams, and under the rails, suspended by bolts from each floor beam. By this arrangement it is evident that the whole weight upon the railroad track is supported by the middle truss.

The upper chord is represented in the details on the plate; it consists of a cast piece in the middle and rolled plates on each side connected by bolts; the rolled plates are without joints. Beneath the chords are placed angle-blocks at the proper intervals of the panels upon which are cast grooves to fit the plates of the upper chord. The vertical rods are in pairs passing through the chords and angle-blocks; they are $1\frac{1}{4}$ in. in diameter.

The lower chord consists of four plates, 4 in. × $\frac{3}{5}$, below which the suspension rods pass, as represented in the plate.

On the lower side of the chord suspension rods pass through a wrought iron plate, the end of which is extended to a sufficient length to give room for two holes to receive the hooked ends of the lateral brace rods.

The braces and counter-braces are equal in size, and the cross-section is in the form of an X.

The dimensions of the flanges are only $\frac{1}{4}$ inch by $2\frac{1}{4}$ inches, and the area of the cross-section $1\frac{1}{2}$ square inches.

The horizontal lateral bracing is by means of rods. They are hooked into holes in the projecting edges of the plates at the bottom of the suspension rods; the 4 rods forming the diagonals of each panel of the lateral bracing unite at the centre, where they pass through a ring 8 inches in diameter, and are secured by nuts on the inside of the ring, which furnishes the means of lateral adjustment.

The planking consists of two courses. 1st course, 2 inch yellow pine, laid longitudinally. 2d course, $\frac{1}{2}$ inch white oak, inclined at angle of 45°.

Bill of Materials for Bridge over Harford Run.

CAST-IRON.

Each truss contains,
32 lineal feet of upper chord cross-section $6\frac{3}{4}$
    square inches   = 2592 cubic in.
30 horizontal counter-braces, each 49 inches
    long, cross-sections $1\frac{1}{8}$ inches   = 1654 "
8 posts over abutments, 36 inches long,
    cross-sections $1\frac{1}{8}$ inches   = 324 "
22 angle-blocks, each 12 cubic inches   = 264 "
2 rollers at ends of truss, 3 inches in di-
    ameter, **8 inches long**   = 112 "

Total cubic inches of cast-iron in one truss = 4946 "
The weight of which at $\frac{1}{4}$ pound **per cubic**
    inch is   = 1237 pounds.

MALLEABLE IRON IN ONE TRUSS.

32 lineal feet of top chords, cross-section 3
    inches   = 1152 cubic in.
32 lineal feet of bottom chords, cross-section
    6 inches   = 2304 cubic in
13 plates under lower chords $7\frac{1}{2} \times 6 \times \frac{3}{4}$ = 439 "
26 suspension rods, $1\frac{1}{4}$ inches diameter, 45
    inches long   = 1440 "
52 **nuts for** suspension **rods,** $2 \times 2 \times 1$ = 208 "

Total malleable iron in one truss = 5543 "
                        Weight in pounds = 1386 pounds.
For the railroad track are required,
32 bolts, **40** inches long, $\frac{3}{4}$ inch diameter, to
    suspend the longitudinal timbers   = 387 cubic in
22 nuts for same, $\frac{3}{4} \times 1\frac{1}{2} \times 1\frac{1}{2}$ = 37 "

                                         424 "
                            Weight 106 pounds.

Each set of lateral braces requires,
12 rods, ¾ inch in diameter, 9½ feet long   602 cubic in
3 rings, 8 inches in diameter, 2 × ¾   108   "
                                             ─────
                                             710   "
                              Weight = 178 pounds.

Wood for the whole Bridge.

44 floor beams of yellow pine, 6 × 12, 17 feet long B. M. 4,488
Floor plank   do.   2 inches   "   4,356
   do.   oak,   1½ inches   "   3,267
64 lineal feet of timber under rails, 10 × 10   "   533
8 cross beams for lateral braces, 5 × 6, 16 feet
    long   "   860
                                                     ──────
                                                     13,504

The weight of which at 3 pounds per foot B. M. = 40,512 lbs.

Recapitulation of Bill of Materials.

Cast-iron in 5 trusses, 1237 pounds each   6,185 lbs.
Malleable iron in 5 trusses, 1386 pounds each, 6,930
Bolts for railroad track   106
  do. for 2 sets of lateral braces, 178 pounds
       each   356
                                                    ──────
                                                    7,392 lbs.
                              Weight of wood   40,512 "
                                                   ───────
                              Total weight = 54,089 "

Estimate of cost.

6185 pounds cast-iron @ 2¼ cents $139 16
7392   "   malleable iron @ 3¼ cents   240 24
13504 feet B. M. timber and plank, average $15
       per M.   203 56
Workmanship, 22 lineal feet @ $13 per foot   416 00
                                              ───────
                              Total cost $997 96

Estimate of **Cost** *of a Single* **Track** *Bridge similar to the above, with* **2** *trusses,* **16½** *feet from centre to centre.*

2474 pounds cast-iron @ 2¼ cents		$ 55 66
2772 " malleable iron @ 3¼ cents		90 09
284 " do. for lateral braces and track @ 3¼ cents		9 23
3400 **feet B. M.** of timber @ $15 per M.		51 00
32 lineal **feet** workmanship @ **$6½**		208 00
	Total cost	$413 98

Average cost per foot, $13.

Calculation.

The middle truss, which **bears the weight of the railroad track**, sustains a much greater load **than either of the others.** As the length of the bridge **is not sufficient for** more than one locomotive, it will be assumed that the greatest load would be 23 tons, or 1533 pounds per foot.

The permanent load will be the weight of the truss and 16½ **feet of** roadway. It will therefore be,

For the truss itself,

cast-iron	1,237	pounds
malleable iron in truss	1,386	"
" " lateral braces, &c.	284	"
Total	2,907	"
For **the roadway,** 3400 feet B. M. @ 3 pounds per foot	10,200	"
Total weight	13,107	"

Or 410 pounds per foot lineal.

Add for the weight of locomotive   46,000   "

Total weight of middle truss and load = 59,107   "

Principle of Calculation.

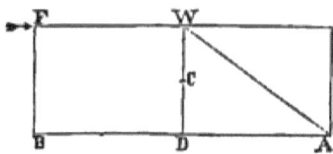

The weight of the half truss AB is supposed concentrated at the centre of gravity C, and acts with a leverage AD equal to one-fourth the span. It is sustained in equilibrium by the horizontal pressure in the middle of the chord, acting with a leverage equal to the height of the truss. The equation of moments is therefore $H \times h = w \times \frac{s}{4}$ or $H \times \frac{w\,s}{4\,h}$,

$H =$ horizontal strain upon the chords,
$w =$ weight on half-span $= 30,000$ lbs. nearly,
$s =$ span $= 30$ feet,
$h =$ height of truss $= 3\frac{1}{2}$ ft.

Therefore $H = \dfrac{30,000 \times 30}{4 \times 3\frac{1}{2}} = 64,286$ lbs. which is nearly the same for the upper and lower chords. To resist this strain we have in the **upper chord $8\frac{1}{4}$ square** inches, and as the strain is compressive it resists with its whole area—the strain is therefore 8000 lbs. **per square inch** nearly.

The lower chord has a resisting cross-section equal to the whole area 6 square inches, for as the span is only 30 feet it is not necessary that there should be a joint.

The strain per square inch will therefore be 10,712 lbs.

The suspension **rods** next **to the** abutments sustain one-half the weight of the bridge, or 30,000 **lbs.**

The cross-section contains $2\frac{1}{2}$ square inches.

The **strain** per square inch with the greatest possible load will be 12,000 lbs.

The strain upon the braces will be to the strain upon the ties in the proportion of the diagonal of the panel to the perpendicular, or in the present case, as 50 is to 42. It is therefore 35,700 lbs. And as the cross section of the two braces 2¼ square inches, the strain per square inch will be 15,800 lbs.

Recapitulation.

Strain per square inch on upper chord   = 8,000 lbs.
   "   "   "   lower chord   = 10,712 "
   "   "   "   end suspension rods = 12,000 "
   "   "   "   braces   = 15,800 "

Maximum load on middle truss 1847 lbs. per square foot.

Weight that would make the greatest strain, 10,000 lbs. per square foot, would be 1169 lbs. per foot.

Breaking weight, if good malleable iron, 60,000 lbs. per square inch.

Greatest strain on bridge = $\frac{4}{15}$ the breaking weight.

LITTLE JUNIATA BRIDGE. (*Plate* 6.)

Description.

The span of this bridge is 60 feet. Its peculiarity consists chiefly in the manner of constructing the arches and the arrangements of the details. The arches are made of iron rails of the U form, such as are frequently used for railroad tracks. Two lines of these rails are placed base to base, breaking joints with each other, and between them is a cast plate with projections on the top and bottom to fit into the cavities of the rails, effectually preventing any lateral separation. The cast pieces are of sufficient length to extend from post to post of each panel, and are varied in size from the middle to the end, so as to be at every point proportioned in cross section to the strain which they are required to bear. This condition renders it necessary that each panel should have a plate of different length and thickness from the next, but as the plates are of a very simple form, and only five patterns are required, the expense is trifling in comparison with the advantage of varying the size according to the pressures; the cast plate is 1 inch by 5 inches in the middle of the span, and $1\frac{1}{2}$ by 5 inches at the ends.

The cross section of the castings, including the projection on the top and bottom, is in the middle   8,214 sq. in.
At the ends   10,714 "
The number of square inches in each rail   4,794 "
The entire cross section of each arch at the middle is   17,802 "
And at the ends   20,302 "

The rise of the arch from under side at skew-back to under side at crown is 8 feet 9 inches.

The upper chords are 3 in number, each 5 × 9, placed $1\frac{1}{4}$ inches apart to allow the diagonal rods to pass between. The length should be 36 feet. The lower chords are 2 in number 6 × 9, placed 5 inches apart, and continuing from pier to pier without joints.

The posts, which extend from the upper to the lower chords, are 4 × 6 inches, and 9 feet 4 inches long from bottom of top chord to top of bottom chord. These posts are in pairs, placed 5 inches apart, to admit of the passage of the arches between them. Between each pair of posts and above the arch is a third post 5 × 8 in. of hard stiff wood, as white-oak, or locust, the office of which is to transmit the pressure of a passing load directly to the arch. The smaller posts serve to connect the system of counter-bracing, give great lateral stiffness to the arch, and, were the failure of the arch under any circumstances possible, they would form the posts of a framed truss sufficient of itself to sustain any ordinary load. There are rods in the direction of both diagonals of the panels, and each set is in pairs.

The rods which extend from the top chords towards the centre, constitute, with the posts just described, a distinct system, which may or may not contribute to any considerable extent in sustaining the load according to the adjustment of the arch. In the absence of the arch they would bear the whole weight. These rods in the first and second panels are $1\frac{1}{4}$ inches diameter, and in the 6 middle panels 1 inch diameter.

The rods in the direction of the other diagonals, or those which beginning at the top chord incline towards the ends of the truss, serve as counter-braces. They have no action in sustaining a direct load uniformly distributed; should the truss settle they would bend, but they have a most important action in resisting any upward pressure, such as is produced by a weight upon one side of the truss, when not counterbalanced by a corresponding weight on the opposite side. These rods are also in pairs, but they are only $\frac{5}{8}$ inch in diameter, except in the last panel, where there is a single rod, the diameter of which is $1\frac{1}{4}$ inches. Plates of annealed copper $\frac{1}{8}$ inch thick are placed in the arches between the ends of the castings to equalize the pressure on the joints.

The width of truss from out to out of chord is 19 feet, and height 10 feet 10 inches.

Between chords in clear   16 ft. 1 inch.

The width of the panels from middle of truss is 6 ft. 1 inch.

The floor beams are 6 × 12 and 24 feet long; between supports the distance is 16 feet 1 inch; they are placed 3 feet from centre to centre, and the weight is distributed by track strings 10 × 10 placed under each rail. These string-pieces are without joints. On the track strings are cross-ties of white oak, or locust, 2 feet apart from centre to centre, 4 × 6 inches in cross-section, laid flat side down.

There are 4 panels of lateral braces to each span.

This system of lateral braces consists of diagonal timbers 5 × 7 resting against angle blocks, and connected by 1 inch bolts extending through both trusses. The same arrangement is used for both top and bottom chords. The system of diagonal braces is represented in the plan. There are two pairs on each pier, and three piers intermediate.

The length of the arch in the middle being 63 feet 8 inches, the most convenient length for the iron rails would be 21 feet 3 inches; by cutting one bar, but not in the middle, there will be pieces to break joints at the ends. The half of **21 feet 3 inches, or 10 feet 7½ inches, would be a convenient length for the castings, and** they would all be of uniform length. The **reason for not** cutting the rail at the middle point is to **prevent** any of the joints from coming opposite to each other. One-third of the length from the end would be **a suitable point of division.**

Bill of Materials for one Span.

PINE.

12	Upper chords	5	×	9	37	ft. long B. M.		1655
4	Lower do.	6	×	9	66	"	"	1188
36	Small posts	4	×	6	9½	"	"	684
23	Floor beams	6	×	12	24	"	"	3312
2	Track strings	10	×	10	66	"	"	1100
6	Purlines	5	×	10	22	"	"	550
6	do.	4	×	6	22	"	"	264
12	do.	3	×	4	22	"	" .	264
2	Guard rails	10	×	12	66	"	"	**1320**
16	Lateral braces	5	×	7	24	"	"	1120
10	Diagonal do.	5	×	6	13	"	"	1950
1300	feet B. M. white pine boards							1300

Total pine = 14707

WHITE OAK.

8	Pier posts	8	×	8	9½	feet long		405
50	Lineal feet posts	5	×	8	8	"		167
34	Cross ties	4	×	6	8	"		544
8	Bolsters	8	×	9	8	"		384

Total white oak B. M. = 1500

CAST-IRON.

18	bottom angle-blocks	wt. each	12 lbs.	216
18	top "	"	25 "	450
4	castings for top chord,	"	50 "	200

63 ft. 8 in. of iron on each side		1808	lbs.		3616
4 skew-backs	wt. each	75	"		300
36 post plates 4 × 6¼ in.	"	5	"	Rolled plates would be preferable.	180
36 do. 13 × 6¼ "	"	20	"		720
16 do. 8 × 8¼ "	"	17	"		272
20 lateral angle-blocks	"	13	"		260
20 washers for lateral bolts	"	3	"		60
208 do. small do.	"	¾	"		174
				Total weight of cast-iron	6448

Malleable Iron for One Span.

63 feet 8 inches of iron rail in each arch, cross-section 9·588 square inches			1832 =	3864	pounds.
16 diagonal rods	1¼ in. diameter,	13 ft. long		868	"
24 do.	1 " "	13 "		827	"
4 counter-brace rods	1¼ " "	13 "		217	"
32 do.	⅝ " "	13 "		432	"
10 lateral bolts	1 " "	19 ft. 6 in.		517	"
60 short bolts through chords ¾ in. diam., 20 in. long				149	"
44 short bolts through posts ¾ " 15 " long				82	"
22 short bolts through top angle-blocks, 1 in. diameter, 9 in. long				44	"
		Total weight		7000	"

Nuts.

20 for 1¼ inch bolts	$1\frac{4}{10}$ lbs.		28 lbs.
66 " 1 "	⅞ "		58 "
104 " ¾ "	½ "		52 "
32 " ⅝ "	½ "		16 "
		Total	154 "
		Weight of rails on bridge	2560 "

Summary.

Weight of timber	per foot	945 lbs.	total weight	56,724 lbs.
do. cast-iron	"	107 "	"	6,448 "
do. malleable iron	"	119 "	"	7,154 "
do. iron rails	"	43 "	"	2,560 "
do. bridge	"	1214 "	"	72,886 "

Estimate of Cost of One Span of Little Juniata Bridge.

14,707 B. M. white pine	$15		$220 60
1,500 " white oak	20		30 00
6,448 lbs. castings	2½ cts.		161 20
7,000 " malleable iron	3½ "		245 00
154 " nuts	9 "		13 86
	Total cost of material		$670 66

Workmanship.

Framing and raising 66 feet, including fitting of arches, at $6½ per lineal foot			$429 00
Making 20 1¼ inch bolts	@ 45 cts.		9 00
" 56 1 "	@ 25 "		14 00
" 104 ¾ "	@ 10 "		10 40
" 32 ⅝ "	@ 10 "		3 20
	Total cost of workmanship		$465 60

Cost of material	per foot	$11 18
do. work, including bolts,	"	7 76
Metal roofing	"	2 50
Painting	"	50
	Total cost per lineal foot	$21 94

LITTLE JUNIATA BRIDGE.

Data for Calculation.

Span	60 feet.
Versed sine	8 " 9 in.
Cross-section of cast plate, in middle of arch	8·214 sq. in.
" " at end of arch	10·714 "
No. of square inches in the section of each rail	4·794 "
Cross-section of arch, at middle	17·802 "
" " ends	20·502 "
" of upper chords, 5 × 9 each,	45·000 "
" " lower " 6 × 9 "	54·000 "
Small posts 4 × 6 "	24·000 "
Middle posts 5 × 8 "	40·000 "
Width of bridge from out to out of chords	19 feet
" between chords in clear	16 "
" panels from middle to middle of post	6 " 1 inch.
Floor beams 6 × 12, 24 feet long, 3 feet from centre to centre.	
Track strings 10 × 10, without joint.	
Cross-ties 4 × 6, 2 feet apart from centre to centre.	
Radius at middle of arch	55·8 feet.
Central angle	65°.
Length of middle line of arch	63 feet 8 inches.
Hypothenuse of skew-back	7·5
Perpendicular	6·32
Base	4·03
Length of rails for arches	21 feet 3 "
Length of cast-segments	10 " 7½ "
Distance from bottom of top chord to top of bottom chord	9 " 4 "
Height from out to out of chords	10 " 10 "
Total weight of bridge without load	72,886 lbs.
Maximum load of 60 lineal feet	120,000 "
Weight of bridge and load	192,886 "
Cross-section of upper chords	270 square inches
" lower chords	180 "

Calculation of Strains.

The calculation as in the case of the Susquehanna bridge, will be made on the hypotheses—

1. That the truss without the arch bears the whole load.
2. That the arch without the truss bears the whole load.
3. That both systems act as one.

FIRST HYPOTHESIS.

That the truss without the arch bears the whole load.

Strain upon the Chords.

To find the position of the neutral axis, the following data are necessary.

Cross-section of upper chords, 270 square inches.
   "   lower chords, 180 " "

Distance from middle **of lower chords to middle of upper, 10 feet 1 inch.**

Let $x =$ distance of neutral axis from middle of the top chord, $(10.08 - x) =$ distance from middle of bottom chord.

Let $P =$ greatest **pressure per square inch** upon **the top** chord. The **pressure being** proportioned to the distance from the neutral axis, **the** pressure per square inch on the lower chord, will be $P \left(\dfrac{10.08 - x}{x}\right)$. The resistance of the top chord will be 270 P. The resistance of the bottom chord, will be 180 $P \left(\dfrac{10.08 - x}{x}\right)$. The weight on one-half of the loaded bridge is 96,433 pounds. The centre of gravity from point of support, **15 feet.**

The equation of equilibrium $270\ P\ x + 180\ P \left(\dfrac{10.08\ x^2}{x}\right)$ $= 96433 \times 15$. Also, $270\ P \times = 180\ P \left(\dfrac{10.08\ x^2}{x}\right)$ from the second of these equations, we find $x = 4.5$ feet, and from the first $P = 595$ pounds $=$ the pressure upon the top chord, and $595 \left(\dfrac{5.55}{4.50}\right) = 610$ pounds per **square inch** $=$ strain upon the **bottom chord.**

LITTLE JUNIATA BRIDGE.

Strain upon the Posts.

In the middle of the bridge the strain upon the posts cannot exceed the greatest load upon one panel, or 19,288 pounds; this is sustained by 4 posts, each 4 × 6, having a united cross-section of 96 square inches. The pressure per square inch in the middle of the span will therefore be 200 pounds.

At the ends of the truss it is proper to calculate the cross-section of the posts at 176 square inches, for if the arches are omitted, the spaces between the small posts must be filled up by extending the end posts to the lower chords.

The weight at one end of the bridge is   9644 pounds
and the pressure per square inch, is   550 "

The formula for the resistance to flexure of the posts is

$$w = \frac{9000\, b\, d^3}{l^2}.$$

We have 4 posts 4 × 6 = 9 feet 4 inches long.
"   2 "   5 × 8 = 9 " 4 "

The weight which would cause the second to yield, is expressed by $w = \dfrac{2 \times 9000 \times 8 \times 5^3}{(9\frac{1}{3})^2} = 206640$,

and for the 4 smaller posts $w = \dfrac{4 \times 9000 \times 6 \times 4^3}{(9\frac{1}{3})^2} = 158693$.

Limit of resistance to flexure = 365,333 pounds.
Greatest weight to cause flexure = 96,433 **pounds.**

Strain upon the Ties.

The strain upon the ties will be to the pressure upon the posts, in the proportion of the diagonal of the panel to the perpendicular, or as 12½ is to 10 nearly, or as 5 : 4, consequently the strain upon the middle ties will be $\dfrac{9288 \times 5}{4} = 24110$ lbs.

The cross-section of the rods is $\cdot 7854 \times 4 = 3\cdot1416$ square inches.

The strain per square inch will be 7680 pounds.

The end ties will bear $\dfrac{96443 \times 5}{4} = 120{,}554$ pounds.

The cross-section of the 4 rods is 5 inches.
The strain per square inch, 24,110.

Lateral and Diagonal Braces.

The amount of side surface is so small, that no doubt can be entertained of the sufficiency of these parts.

(See Calculation on Susquehanna Bridge.)

Floor Beams.

Allowing the heaviest locomotives to have 18 tons on 6 drivers, and the space on the rails occupied by 3 pair of drivers to be 11 feet, the weight may be considered as equally distributed on 5 floor beams, which would give $3\frac{3}{5}$ tons to each.

This weight acts at a distance of 2 feet 5 inches from the centre of the beam, and as the floor beams are 16 feet 1 inch between supports, a weight of $3\frac{3}{5}$ tons, at a distance of 5 feet $7\frac{1}{2}$ inches from the point of support, will be equivalent to $\dfrac{3\frac{3}{5} \times 5\frac{7\frac{1}{2}}{12}}{8}$ applied in the middle $= 2\frac{17}{32}$ tons, or 5066 pounds.

To this must be added one-half of the weight of the beam itself $\dfrac{24 \times 6 \times 3}{2} = 216$ pounds, and the total weight in the middle of the beam will be 5282 pounds.

The formula for the strength of a beam supported at the ends, is $R = \dfrac{18\, w\, l}{b\, d^2}$ where l is in feet,

Therefore, $R = \dfrac{18 \times 5282 \times 16}{6 \times 12^2} = 1760$ pounds $=$ maximum strain per square inch.

The deflection caused by the passage of a locomotive with 18 tons weight upon the drivers, will be deduced from the equation $w = \dfrac{BD^3}{.0125\, l^2} = \dfrac{6 \times 12}{.0125 \times 16^2} = 3240$ pounds weight, that will cause a deflection of $\frac{1}{40}$ inch to 1 foot, or $\frac{16}{40} = \frac{2}{5}$ of an inch in 16 feet.

The actual weight being 5282 pounds, the deflection will be in proportion, or $\frac{2}{5} \times \dfrac{5282}{2240} = .65$ inch deflection caused by the passage of a locomotive.

Counter-Braces.

The greatest possible strain upon the counter-braces, being equal to the strain upon the braces of the middle panels due to the variable load, will be 1200 pounds.

The cross-section of the 4 rods ⅝ diameter is 1¼ square inches. The greatest possible strain per square inch, 9600 pounds.

SECOND HYPOTHESIS.

Calculation of the strength, on the supposition that the arch supports the whole weight.

The span of the arch is 60 feet, and rise 8 feet 9 inches.
The weight on the half arch being 96,443 pounds.
Distance of centre of gravity from support, 15 feet.
Cross-section of two arches in middle, 35·6 square inches.
Cross-section of two arches at ends, 40·6 do.
$w = P =$ pressure per square inch, we will have
$$P \times 35.6 \times 8.75 = 96443 \times 15,$$
whence $P = 4644$ pounds = strain per square inch—middle of arch.

The pressure at the skew-back is to the pressure at the crown as the hypothenuse is to the perpendicular, or as 7·50 : 6·32.

But the cross-section at the skew-back is also increased in

the proportion of 20·3 to 17·8. The pressure at the skew-back will, therefore, be per square inch

$$4644 \times \frac{7·50}{6·32} \times \frac{17·8}{20·3} = 4832 \text{ pounds.}$$

Strain *upon* the Counter-Braces.

(See figure used in calculating Susquehanna Bridge.)

The greatest variable load on one half the bridge is 60,000 pounds.

We will have in this case
$BC = 15$ feet.
$CG = \frac{3}{4} \times 8·75 = 6·50$ "
$AG = \sqrt{45^2 + 6·56^2} = 45·5$ "
$GD = 11·4$ "
$fG = 4·9$ "
$GB = 9·8$ "

Resultant in the direction $GA \dfrac{60000 \times 11·4}{6·56} = 104270,$

and $\dfrac{104270 \times 9·8}{22·5} = 50000$ nearly = maximum limit of the upward pressure upon the arch.

The strain per square inch upon the **counter-brace rods of one panel** resulting from the pressure will be 10,000 pounds nearly.

THIRD HYPOTHESIS.

Calculation of the strain, on the supposition that both systems act as one. As the arch thrusts against a skew-back placed upon and **between** the **bottom** chords, it is important to inquire **whether the** whole **strain is** sustained by **the** lower chord, **or whether any** assistance is derived from the masonry itself.

Where the roadway of a bridge is placed upon the bottom **chord, the** chords generally rest upon the tops of the abutments, **and if the arch is** attached to the chord, as in the plan now **under consideration, it is evident that the** latter must bear the whole strain, and no assistance whatever can be derived from the resistance of the masonry. When the roadway is on the top chord, the masonry is usually carried up to the level of the road with an offset for the bottom chord to rest upon. And it is never necessary in this case that the lower chord should bear the whole strain. By placing a wall-plate behind the ends of the lower chords, and driving wedges between it and the chords, a pressure is thrown upon the abutment, which takes off precisely an equal amount from the strain upon the chord. The assistance to be derived from this arrangement is very great, and should never be neglected where circumstances admit of its being employed. It is not safe to depend entirely upon the resistance of the abutment, or, in continuous spans, upon the counterbalancing thrust of one arch against the next, for the loss of one span in this case would insure the destruction of the whole.

It is seldom, however, that when one span of a bridge is carried away the next to it is loaded with much more than its own weight, and, consequently, the true minimum of the size of the lower chords should be such as would render it more than sufficient to sustain the tension arising from the weight of the bridge. It will be safe to depend upon the mutual assistance of the spans and of the abutments to sustain the greater proportion of the thrust arising from the variable load. In the present case, as the spans are so short that the lower chord can be made without joints, there is a greater resisting power than is required to sustain the loaded bridge, since we have seen that the strain was less than 600 pounds per square inch.

The present calculation will be made upon the supposition that the chords are keyed at the ends next to the abutments, and in close contact over the pins, which is equivalent to doubling the resisting area. The following are the data for calculation in this case:

From middle of upper to middle of lower chord, 10 feet.
From middle of upper chord to middle of arch, 8 feet.
Middle of skew-back and middle of lower chord on same horizontal line.

As the arch does not abut against the masonry, but rests **upon and between** the lower chords at the skew-back, **the limits** of resistance in a horizontal direction will be the resistance of the cross-section of the chords, for if the thrust should be greater than this, whatever may be the strength of the arch, the ends of the chords would be crushed. Consequently, the total resistance of the arch and lower chords on the line **of the** skew-back, must be equal to twice the resistance of the cross-section of the chords, or to 400 square inches.

After making allowances for bolt-holes, &c., the cross-section of the upper chord is 270 square **inches.**

The **cross-section** of the arch is 35·6 square inches, and **allowing the** resistance of iron to be ten times as great as that **of wood, the equivalent** cross-section would be 356 square inches.

Let $x =$ **distance of neutral axis** from middle of **top chord;** $(x - ·66) =$ **distance from** middle of arch; $(10 - x) =$ **distance** from middle of bottom chord.

Let $P =$ **maximum pressure per square inch on top chord.**

$\dfrac{P}{x}(x - ·66) =$ **pressure** per square inch upon arch.

$\dfrac{P}{x}(10 - x) =$ pressure per square inch on bottom chord.

The equations of equilibrium are,

$$270\,P\,x + 356\,\frac{P}{x}(x - ·66)^2 = 400\,\frac{P}{x}(10 - x)^2 \text{ and}$$

$$2 \times 400\,\frac{P}{x}(10 - x)^2 = 96443 \times 15.$$

From the **first** equation we find $x = 4·64$.
And from **the** second $P = $ **292** pounds.

The strain **per** square inch **on the** upper chord being 292 **pounds.**

On arch at crown it will be $\dfrac{292 \times 4}{4·64} \times 10 = 2517$ pounds.

On the lower chord it will be $\dfrac{292 \times 5·36}{4·64} = 337$ pounds.

Vertical *Pressure upon* the Arch and Posts.

The pressure of the arch in the direction of the tangent at the skew-back may be resolved into two components, one horizontal, and the other vertical; these will be proportioned to the perpendicular and base of a right-angled triangle, of which the face of the skew-back is the hypothenuse; and if the length of the hypothenuse be taken to represent the thrust of the arch, the base will represent the vertical pressure or portion of the weight sustained at that point.

The section of the arches at the ends is   40·6.
The hypothenuse of the skew-back   7·50.
The base of skew-back   4·03.

The proportion of surface which resists the vertical pressure $\frac{40\cdot6 \times 4\cdot03}{7\cdot50} = 21\cdot8$ of iron, equivalent to 218 square inches of wood. The cross-section of 4 posts, each $4 \times 6 = 96$ inches.

Total resisting surface   314 square inches.
The weight being   96,443 pounds.
The pressure per square inch is   307 "

In the middle, the maximum pressure upon the posts can never exceed the amount previously determined as due to the weight upon one panel.

The posts at the ends containing 96 square inches, and the pressure per square inch being 307 lbs., the proportion of the weight sustained by the truss will be 29,472 lbs., which produces a pressure in the direction of the diagonal rods $\frac{29472 \times 5}{4} = 36,840$ lbs. As the cross-section of the four rods is 5 square inches, the strain per square inch will be 7,368 lbs. The strain upon the counter-braces will be the same as in the other cases.

General Summary of Results.

No. of feet B. M. white-pine in one span   14,707
"   "   white-oak   "   "   1,500

No. of pounds of cast-iron — 6,448
" " rolled-iron — 7,000
" " nuts — 154
Weight of timber per lineal foot — 945
" cast-iron " " — 107
" rolled-iron " " — 119
" nuts " " — 43
" finished bridge per lineal foot — 1,214
" bolts for arches per foot — 115
Cost of workmanship of one span — $466 00
" material " " — $671 00
Total cost per lineal foot — $22 00
If the truss be supposed to bear the whole load,
The pressure upon the top chord will be — 595 pounds.
" " " bottom chord will be — 610 "
" " " posts in the middle of span — 200 "
" " " a post at end of span — 550 "
" " " the ties in the middle — 7,680 "
" " " the ties at the ends — 24,110 "
Maximum load upon the floor-beam — 3⅜ tons.
Equivalent weight in middle — 5,282 pounds.
Maximum strain per square inch — 1,760 "
Deflection of floor-beam by weight of locomotive — ·65 inches.
Greatest possible strain per square inch, counter-braces — 9,600 pounds
If the arches bear the load, the strain will be
in the middle of arch, per square inch — 4,644 "
At the ends of the arch, per square inch — 4,832 "
Upon the counter-braces — 10,000 "
If arches and truss act together as one system,
The maximum pressure of top chord per square inch — 292 "
The maximum pressure of bottom chord per square inch — 337 "
The maximum pressure of the arch per sq. inch — 2,517 "
Pressure per square inch on posts at end of span — 307 "
" " " " in middle — 200 "
Pressure on ties at end of span — 7,368 "
" " middle of span — 7,680 "

SHERMAN'S CREEK BRIDGE—PENN. CENTRAL RAILROAD. (*Plate* 7.)

This structure, in the general appearance of the elevation of the side-truss, bears some resemblance to a Burr Bridge, but it possesses several peculiarities.

1. The truss is double, consisting of three rows of top and bottom chords, and two sets of posts and braces.
2. The truss is counter-braced by inch rods, placed between the braces, and running in nearly a parallel direction. These rods pass through bolster-pieces, placed behind the posts on the top and bottom chords.
3. The panels increase in width from the ends towards the middle of the spans. The first panels are 9 feet $1\frac{1}{2}$ inches from centre to centre of posts.

The middle panels 12 feet $1\frac{1}{2}$ inches.

The bridge consists of 2 spans, each 148 feet 3 inches from skew-back to skew-back, or 154 feet 6 inches from middle of pier to end of truss. The pier is 3 feet 2 inches on top, and 6 feet at skew-backs.

The foundation of the pier presented some peculiarities in its mode of construction. Great difficulties were apprehended in consequence of an opinion, based upon information given by residents in the vicinity of the work, that the rock was at a great depth, and was covered by a deposit consisting of the ruins of an old dam. As the rock could not be reached by sounding, before the excavations were commenced, in consequence of the large stones which were scattered through the gravel, it was concluded to make use of a crib, consisting of timbers solidly and compactly framed together without leaving space between. The timbers of one course lie in immediate contact with those of the next, and the whole are bolted together with iron rods. The intention was to make use of this as the frame of a coffer-dam, if it was found possible to reach the rock, and keep out the water; if not, to use it as an ordinary crib, and fill it with rough stones.

The process of excavating the foundation was commenced with a horse-power dredging-machine, consisting simply of a scoop, capable of holding about 6 cubic feet, with handles at front and rear by means of which it could be held down by four or eight men. The point of the scoop was shod with iron, and armed with teeth, a chain was attached to the point passing round a windlass, to which a horse was attached. With this simple apparatus, the bottom was excavated to the surface of the rock in a few days.

The crib was sunk, puddled on the outside, and the water bailed. It was found to answer effectually as a coffer-dam. The masonry was carried in regular courses to the surface of the water, the space between the regular masonry and the crib filled with stones, and the whole grouted perfectly tight with hydraulic cement. The whole expense of the foundation was $400 — including excavation with machine, bailing, puddling, and grouting.

Bill of Timber for one Span of Sherman's Creek Bridge.

3 wall-plates	8 × 16	18 feet long B. M.	576
20 chords	6 × 13	36 " "	4,680
10 "	8 × 13	36 " "	3,120
10 "	8 × 10	36 " "	2,400
20 "	6 × 10	36 " "	3,600
56 posts yellow-pine	9 × 12	23 " "	11,592
4 king-posts	9 × 16	23 " "	1,104
15 floor-beams	8 × 14	18 " "	2,520
14 "	7 × 14	18 " "	2,058
56 lateral braces	$4\frac{1}{2}$ × 7	$8\frac{1}{4}$ " "	1,213
3 " "	$4\frac{1}{2}$ × 7	13 " "	103
30 roof-braces	4 × 5	17 " "	850
56 check-braces	9 × 20	3 " "	2,520
56 "	9 × 23	3 " "	2,898
60 main-braces	6 × 9	19 " "	5,130
15 tie-beams	8 × 10	19 " "	1,900

Amount carried over 46,264

SHERMAN'S CREEK BRIDGE.

			Amount brought forward		46,264
8 purlines	4 × 6	20 feet long B. M.			320
135 rafters	3 × 5	10½	"	"	1,772
15 roof posts	4 × 5	3	"	"	75
30 knee-braces	5 × 5	5	"	"	312
16 track stringers	8 × 10	20	"	"	2,133
8300 feet B. M. ¾ inch sheeting for roof					3,300
56 arch-pieces	9 × 11	25	"	"	11,550
7000 feet B. M. inch boards, for weather-boarding, 20 feet long					7,000

72,726

Weight per lineal foot, 1,416 pounds.
No. of cubic feet per foot lineal, 40.

Bill of Counter-Brace Rods for one Span.

4 rods for	1st panels each	24 ft. 3 in. long	1 in. diam.	97. ft.
4 "	2d " "	24 " 2 "	1 "	97. "
4 "	3d " "	24 " 8 "	1 "	98·7 "
4 "	4th " "	25 " 0 "	1 "	100. "
4 "	5th " "	25 " 2 "	1 "	100·7 "
4 "	6th " "	25 " 8 "	1 "	102·7 "
4 "	7th or middle pan.	26 " 0 "	1 "	104·0 "

Total lineal feet 700

Weight in pounds at $2\frac{65}{100}$ per foot = 1,855 pounds.

Arch Suspension Rods for one Span.

4 rods each 6 feet 8 inches ⎫
8 " 10 " ⎪
8 " 13 " ⎬ $1\frac{3}{8}$ inches diameter.
8 " 15 feet 6 inches ⎪
8 " 17 " 2 " ⎪
8 " 18 " 2 " ⎭

Total length 322 feet.
Weight at $4\frac{94}{100}$ lbs. per foot, 1,590 pounds.

Lateral Brace Rods for one Span.

15 rods each 16 feet 9 inches long 1 inch diameter 655 pounds

Small Bolts for one Span.

60 bolts, through arches, 47 inches long 1 inch diam. 622 lbs
60 bolts, through chords and posts, 34 inches long $\frac{3}{4}$
    inch diam.   255 "
30 roof-bolts 36 inches long $\frac{3}{4}$ inch diam.   135 "
224 spikes for braces $\frac{3}{4}$ pound each   168 "

Dimensions and Data for Calculation of Bridge at Sherman's Creek.

Span at skew-backs	148 ft. 3 in.
Whole length of truss for one span	154 "
Out to out of chords	20 "
Middle to middle of chords	19 "
Resisting cross-section of upper chords	400 sq. in.
Resisting cross-section of 6 lower chords, deductions for splice, check-brace and bolt, and allowing for scarf-key	280 sq. in.
Versed sine of lower arch	20 feet.
Cross-section of 8 arches	800 sq. in.
Span 148$\frac{1}{4}$, and rise 20, will give radius	172·25 feet
And 172·25, 152·25, and 74·125, express the proportion of the hypothenuse, perpendicular, and base of skew-back.	
Hypothenuse of skew-back covered by arches	18 inches.
Perpendicular " " "	16 "
Base " " "	7·6 "
Distance from skew-back to bottom of chord	4$\frac{1}{2}$ "
" " middle of skew-back to middle of chord	4 ft. 5 in.
Width from out to out of chords	16 " 2 "
" between chords in the clear	11 "
Distance from centre to centre of floor-beams	5$\frac{1}{2}$ feet.

Weight of one-half span complete (77 feet)   120,000 lbs.
Distance of centre of gravity from point of support 37 feet.
Weight of one-half span with load   275,000 lbs.
Distance between shoulder of post   15½ feet.

Calculation of Truss without the Arches.

Let x = distance of neutral axis from top chord.
$19 - x$ = distance from bottom chord.
P = pressure per square inch on top chord.
$\dfrac{P}{x}(19-x)$ = strain per square inch on bottom chord.

$400\, P\, x = 280\, \dfrac{P}{x}(19-x)^2.$

$x = 8\cdot 3$ = distance from top chord.
And $19 - x = 10\cdot 7$ = distance from bottom chords.

$400\, P \times 8\cdot 3 + 280\, P \times \dfrac{10\cdot 7}{8\cdot 3} \times 10\cdot 7 = 275{,}000 \times 37.$

$P = 1532$ lbs. = pressure per square inch on top chord.

And $1532 \times \dfrac{10\cdot 7}{8\cdot 3} = 1{,}975$ lbs. = strain upon bottom chord.

The bottom chords derive some assistance from the masonry, but as the roadway is on the bottom of the truss, little opportunity is given for wedging the lower chords, and for this reason the assistance to be derived from this service is not estimated.

Ties and Braces.

The weight upon the middle panel (12¼ lineal feet) is 45,000 lbs. To resist this there are four posts, the cross-section of each being 72 square inches, or the united cross-section 288, equivalent to 156 lbs. per square inch.

The distance between the shoulders of the posts being 15½ feet, and the width of the middle panel, exclusive of posts, 11½ feet, the diagonal will be 19·3.

The strain upon the diagonal will be $45{,}000 \times \dfrac{19\cdot 3}{15\cdot 5} = 56{,}000$

lbs., which divided by the cross-section of the four braces, will make the pressure per square inch $\frac{56,000}{216} = 260$ lbs.

The expression for the limit of the resistance to flexure $w = 4 \times \frac{9,000 \, B \, D^3}{l^2}$ gives the present case $w = \frac{9,000 \times 9 \times 6^3}{19 \cdot 3^2} =$ 46,000 pounds, or for

The four braces   184,000 pounds.
The actual pressure   56,000   "
Difference in favor of stability 128,000 "

The end ties sustaining the weight of half the bridge, will be at 275,000 pounds, the cross-section being as before 288 square inches, the strain per square inch will be 955 pounds.

The width of the end panel being 8½ feet exclusive of posts and the distance between the shoulders of the posts being as before 15½ feet. The diagonal will be 17·7 feet, and the pressure in the direction of the braces $\frac{275,000 \times 17 \cdot 7}{15 \cdot 5} = 314,000$ pounds = 1451 pounds per square inch.

The limit of the resistance to flexure for the 4 braces is expressed by $w = \frac{9,000 \times 9 \times 6^3}{17 \cdot 72} \times 4 = 223,000$ pounds.

As the pressure is 314,000 pounds, it appears that with the assumed weight of a train of locomotives, or one ton per lineal foot besides the weight of the structure, the end braces would yield by lateral flexure in the direction of the plane of the truss if not supported in the middle.

If an intermediate support be used, the resistance will be quadrupled, and will be amply sufficient.

It is also necessary to examine whether the braces, if supported in the middle in the direction of the plane of truss, could yield laterally in the direction of the perpendicular to this plane; the relative resistance in the two cases, are as 6×9^3 : 9×6^3, or as 9 : 4. The limit in this case would therefore be $\frac{223,000 \times 9}{4} = 502,000$ pounds, which is more than the pressure (314,000 pounds).

It appears therefore from this calculation, that if the arches are omitted, the end braces should be supported in the middle by diagonals in the opposite direction. As an additional security, the depth should be increased to 9 inches. In the other panels they should diminish gradually to the middle of the span, where the original dimensions are sufficient.

Floor Beams.

The floor beams are 7 × 14 inches, width in clear between supports 11 feet, distances from centre to centre $5\frac{1}{2}$ feet.

The weight on the drivers of a locomotive 18 tons, may be considered as distributed nearly equally over 3 floor beams, which will give 6 tons for each beam.

$6 \times 3 \div 5\cdot5 = 3\cdot3$ tons = the equivalent weight in the middle of the beam

$$R = \frac{18\,w\,l}{b\,d^2} = \frac{18 \times 6600 \times 11}{7 \times 14^2} = 952 \text{ pounds} = \text{maximum}$$

strain per square inch.

Lateral Braces.

The lateral braces are $4\frac{1}{2} \times 7$ and 8 feet long. The prevalent winds are usually in a direction nearly parallel to the axes of the bridge, so that its exposure is not great. Assume as the basis of a calculation that the sides are closely boarded 20 feet high, and that the perpendicular force of wind may be 15 pounds per square foot, the whole pressure upon one span will be 45,000 pounds. As there is lateral bracing both above and below, this pressure would be resisted by 4 lateral rods 1 inch diameter = 3·14 square inches, or 3,344 pounds per square inch.

The proportional strain upon the lateral braces would be $\frac{45,000 \times 8}{5} = 72,000$, to resist which, are 4 braces $4\frac{1}{2} \times 7 = 126$ square inches = 571 pounds per square inch. The bearing surface at the joints does not much exceed one-half the area

15

of the cross-section, consequently the actual pressure at the joints will be about 1,000 pounds.

The limit of flexure of the 4 braces, is expressed by $w = \dfrac{9,000 \times 7 \times 4\frac{1}{2}}{8^2} \times 4 = 360,000$ pounds nearly.

The maximum pressure is 72,000 pounds.

Difference in favor of stability, 128,000 pounds.

The lateral braces cannot yield either **by** crushing or bending, **and** are, therefore, amply sufficient.

Could the bridge, if not loaded, be blown away?

The weight of one span has been found to be **240,000** pounds.

The resistance to sliding would be   120,000 pounds.
The pressure of wind   45,000   "

Difference in favor of stability   75,000   "

Could the *bridge yield* **to** *the force* ***of the wind by*** *rotation around the outer edge of the chord?*

The effect of the wind, 45,000 pounds, **acting** with a leverage of **10 feet, would give** for **the disturbing force**   450,000 pounds.

The **resistance =** weight of bridge × half-width from out to out = 240,000 × 8 = **1,920,000**   "

Difference in favor **of** stability   1,470,000   "

Strain upon the Knee-Braces

This is **the** first case in which it has been necessary to calculate the **strain** upon a knee-brace. The cross-bracing of **the** other bridges upon which calculations have been made **having** been in the direction of **the** diagonals.

The case, however, presents no difficulty. Let $A\ C\ B\ D$ **represent** the cross-section. The effect of the pressure of wind on $A\ C$ is equivalent to half that pressure applied at the point A. A **force** at A tends to produce rotation around B and C, **which may be** resisted by a brace in the direction of the diagonal $A\ B$.

SHERMAN'S CREEK BRIDGE.

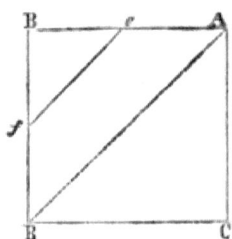

The pressure upon the brace will bear to the force at A the proportion of the diagonal to the side $A\,D$. If the brace be removed the pressure must, nevertheless, still continue, and if it is resisted by a brace $e\,f$, the pressure upon $e\,f$ will be greater than upon $A\,B$ in the proportion of $A\,D$ to $e\,D$, because D is a fulcrum and $A\,D$ and $e\,D$ the leverages of the acting and resisting forces. If $e\,f$ is parallel to $A\,B$, which is generally a very favorable direction, the length $e\,f$ and $A\,B$ will be in proportion to the distances $D\,e$ and $D\,A$, and may be substituted for them. In the present case the force of wind, 45,000 pounds, acting with a leverage of ten feet, will give its moment 450,000, or 225,000 pounds acting at a distance of 20 feet. The length of the diagonal is $\sqrt{20^2+16^2}=25{\cdot}6$ feet. and the strain in the direction of the diagonal $\dfrac{22500\times25{\cdot}6}{16}$ = 36,000 pounds.

The length of the knee-braces being 5 feet, the strain upon them will be $36{,}000\times\dfrac{25{\cdot}6}{5}=184{,}000$ pounds. This is resisted by 15 braces (one to each post). The cross-section of each is 25 square inches, but, as the bearing surface of the joint does not extend over the whole surface of the section, the resisting portion will be reduced to 15 square inches. The strain per square inch will therefore be $\dfrac{184000}{15\times15}=818$ pounds

For the resistance to flexure of the 15 braces, $w=\dfrac{9000\times5\times5^3}{5^2}\times15=3{,}555{,}000$, or about 20 times the pressure.

The strain upon the bolts at D, will be to the vertical component at A, in the proportion of DE to EA, or as $5 : (25 \cdot 6 - 5)$. The vertical component at $A = 22{,}500 = \dfrac{AD}{AC} = 22{,}500 \times \dfrac{16}{20} = 17{,}000$. Hence, the strain upon the 15 bolts will be $17{,}000 \times 4 = 68{,}000$, or $4{,}533$ pounds to each bolt, or $10{,}000$ pounds per square inch if the bolts are $\frac{3}{4}$ inch diameter.

Pressure upon the Arch.

For this calculation we have, from the table of data,
Span, 148 feet.
Distance of centre of gravity from abutment, 37 feet.
Rise of arch, 20 feet.
Proportion of hypothenuse, base, and perpendicular of skew back = 18 7·6 and 16.
Cross-section of 8 **arches, 800 square inches.**

$800 \times \dfrac{16}{18} = 711$ proportion **to** resist horizontal thrust **at** skew-back.

$800 \times \dfrac{7 \cdot 6}{18} = 338$ **square inches to** resist vertical pressure at skew-back.

The weight for one-half span loaded, is 275,000.
$$800 \times 20 \times P = 275{,}000 \times 37.$$
$P = 448 =$ **pressure** per square inch, **on** arches in middle.

The resisting cross-section at **the** skew-backs is the same as at the crown.

The **pressure is** greater in **proportion** of the hypothenuse to the perpendicular; it will therefore be $448 \times \dfrac{18}{16} = 504$ lbs.

The arches are therefore more **than** sufficient to sustain the whole weight.

When both systems act as one.

The data **required to** determine the **strains** upon the chords **and** arches are,

SHERMAN'S CREEK BRIDGE.

Distance from middle of upper to middle of lower chord .... 19 feet.
Distance from middle of skew-back to middle of lower chord .... 4⅛ "
Distance from middle of top chord to middle of arch .... 3·5 "
Cross-section of upper chords .... 400 square in.
  "   lower   "   280   "
  "   arch at crown 800   "
  "   " skew-backs ... 711   "

Let x = distance of top chord from neutral axis.
" $x - 3·5$ = distance of arch at crown from neutral axis
" $19 - x$ =   " bottom chord   "   "
" $23·5 - x$ =   " arch at skew-back   "   "
" P = pressure per square inch, on top chord.
" $\dfrac{P}{x}(x - 3·5) =$ "   arch at crown.
" $\dfrac{P}{x}(19 - x) =$ "   bottom chord.
" $\dfrac{P}{x}(23·5 - x) =$ "   arch at skew-back, horizontally.

The equations in this case are,

$$400\,P\,x + \dfrac{P}{x}800\,(x-3·5)^2 + \dfrac{P}{x}280\,(19-x)^2 + \dfrac{P}{x}711\,(23·5-x)^2 = 275{,}000 \times 37,$$

and $400\,P\,x + 800\,\dfrac{P}{x}(x-3·5)^2 = 280\,\dfrac{P}{x}(19-x)^2 + 711\,\dfrac{P}{x}(23·5-x)^2.$

From the second of these we find $x = 11·8$.
Consequently the distance of the neutral axis will be,

Below top chord .... 11·8 feet
  " arch .... 8·3 "
Above bottom chord .... 7·2 "
  " skew-back .... 11·7 "

These values substituted in the first equation will give $P \times 222{,}000 = 7{,}175{,}000 \times 11·8$, or

$P = 381$ lbs. = pressure per square inch on top chord.

$381 \times \dfrac{8\cdot 3}{11\cdot 8} = 268$ lbs. = pressure per square inch on arch at top.

$381 \times \dfrac{7\cdot 2}{11\cdot 8} = 232$ lbs. = strain per square inch on lower chord.

$381 \times \dfrac{11\cdot 7}{11\cdot 8} = 377$ lbs. = strain per square inch on perpendicular of arch at skew-back.

Vertical Pressure.

Assuming that the weight sustained by each system will be in proportion to its power of resistance, the greatest weight that the truss can sustain will be the limit of flexure of the braces in the end panels. This has already been found to be 223,000 pounds, which will be produced by a vertical pressure of $\dfrac{223{,}000 \times 15\cdot 5}{17\cdot 7} = 200{,}000$ pounds: this is the extreme limit of the power of resistance of the end braces.

The proportion of surface at the skew-back which resists the vertical pressure is 388 square inches. If we suppose the vertical pressure on the base of the skew-back to be the same per square inch as the horizontal pressure upon the perpendicular, it will be capable of resisting 180,830 pounds; this deducted from the whole pressure, 275,000 "

will leave for the portion to be sustained by
the braces 94·170 "
which is below the resisting power. The actual limit of the resisting power of the arch is very great, but, assuming that in practice it is not safe to exceed 1000 pounds per square inch, the proportions of the weight sustained by the truss and arch would be,

For the truss $275{,}000 \times \dfrac{108{,}663}{338{,}000 + 108{,}663} = 68{,}700$ nearly.

And for the arch $275{,}000 \times \dfrac{338{,}000}{446{,}663} = 206{,}100$ nearly.

These numbers will give for the strain per square inch on the arch, $\frac{206,100}{338} = 600$ lbs. nearly.

For the end braces $\frac{68,700 \times 17\cdot7}{15\cdot5 \times 216} = 360$ lbs. nearly.

It has been stated that the bridge at the western end is sustained by an abutment pier—it is proper to examine whether the resistance which it is capable of opposing is sufficient to counterbalance the thrust of the arch, on the supposition that it should bear the whole of the load.

The dimensions of the abutment pier are given in the following figure, except the length, which may be taken at 16 feet.

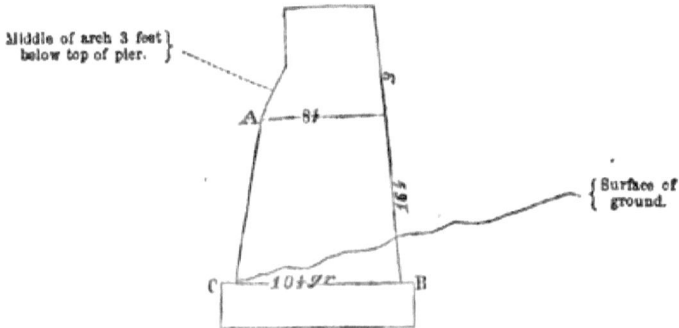

We will examine the conditions of equilibrium on the supposition that rotation takes place around the point B. The disturbing force is the horizontal component of the thrust of the arch $= 358,750$ lbs. acting with a leverage of $16\frac{1}{2}$ feet, its moment will therefore be $358,750 \times 16\frac{1}{2} = 5,919,375$.

The resistances are,

1. The weight of the masonry above $CB = 110$ perches of 3,750 lbs. $= 412,500$ lbs. The distance of centre of gravity from B is 5 feet, the moment will be 2,062,500.

2. The adhesion of the mortar, estimating it at 50 lbs. per square inch, or one-half the tabular strength of hydraulic cement, will be on a surface of 160 square feet $= 1,152,000$ lbs., and its moment with a leverage of 5 feet $= 5,760,000$ lbs.

3. The vertical pressure of the arch itself, 275,000 lbs.,

acting with a leverage of 9 feet, will give a moment 275,000 × 9 = 2,475,000.

The sum of the moments of the resisting forces will be 2,062,500
5,760,000
2,475,000

Total 10,297,500
Moment of disturbing force 5,919,375

Difference in favor of stability = 4,378,125

As this difference is less than the **adhesion** of the mortar, it appears that an **abutment** pier of **dry masonry** of the same dimensions would be overturned.

It has been supposed in this calculation that the arch bears the whole weight, and that the abutment resists the whole thrust. The actual horizontal thrust, with the two systems acting together, was found to be 377 × 711 = 268,047. The moment will be 268,047 × 16½ = 4,422,775. The resistance, omitting the strength of the mortar = 4,537,500. From which it appears that if we disregard the adhesion of the mortar, the system as a whole would be very nearly in a state of equilibrium, the difference being in favor of stability. The practice of the writer in proportioning abutments on **rock** foundations is, to disregard the adhesion of the mortar, throwing this, whatever it may be, in favor of stability; there is so little uniformity in the strength of mortar, and so much liability to cracks occasioned by jars, when partially set, that it is not safe to depend too much upon it. If the proportions and weight of an **abutment prevent** it from overturning, without taking the **strength of the mortar** into consideration, it **is too weak.**

When the base is to any extent compressible, it is not sufficient that the disturbing and resisting forces should be in a state of equilibrium, a condition which requires the resultant of all the forces to pass through the point of rotation. But it is proper that the resultant should pass through the middle of the base.*

---

* This calculation was made before the completion of the bridge; the correctness of the conclusions was soon confirmed; the pier began to crack after the opening of the road, and an increase of thickness by the addition of buttresses was found necessary.

Estimate of Cost of One Span.

72,726 feet B. M. timber	12½ cts.	$909 07
5,280 lbs. rolled iron	3½ "	184 80
3,300 square feet, roof	10 "	330 00
Making 147 bolts	30 "	44 10
" 90 "	10 "	9 00
Workmanship of 154 feet at	8 "	12 32
		$2,708 97

Cost per foot, $18 24.

Summary.

Span	148 ft. 3 in
Width of pier on top	3 " 2 "
" " skew-back	6 "
Timber in one span	72,726 "
Weight of timber per lineal foot	1,416 pounds
No. of cubic feet per foot lineal	40 "
Weight of iron in one span	5,280 "
Width from out to out of chords	20 feet.
" middle to middle of chords	19 "
Versed sine of lower arch	20 "
Radius	172,55 "
Weight of half-span loaded	275,000 pounds
Strain upon floor beams per square inch	902 "
" lateral brace-rods per square inch	3,444 "
" lateral braces	571 "
" knee-braces per square inch	818 "
Pressure per square inch on top chord	381 "
" " " arch at crown	268 "
" " " lower chord	232 "
" " " arch at skew-back	600 "
" " " end-braces	360 "
" " " middle braces	260 "

RIDER'S PATENT IRON BRIDGE. (*Plate* 8)

Description.

The truss of the Rider bridge is principally composed of an upper and lower chord, upright posts, and diagonal ties. The upper chord is **made of cast iron, with** heavy horizontal flanges. The **lower chord is made of** wrought-iron. The diagonal ties, **or suspension rods,** are also made of wrought-iron, and **are secured only to the** upper and lower chord, at **regular intervals, running upwards and downwards** diagonally with the chords, and at nearly right angles with each other. The posts are of cast-iron, and are placed at equal distances apart along the whole length of the chords, to keep the upper and lower chords asunder, and at the same time to assist in preserving the truss in line.

A wedge is inserted on the top of each iron post, under the top chord, by the action of which the diagonal rods are kept in a state of tension.

Bill of Materials for a *Single Span of* 60 *feet.*

Height of truss, **7 feet.** Width in the **clear of** chords, 12 feet. Floor-beams **of wood.**

Cast-Iron.

54 cast-iron posts 6 feet long, cross-section 9 sq. in. 8,784 lbs.
132 lineal feet **cast-iron** top chord, "   15   "   5,940 "
110   "   caps for lower chord,   "   2½   "   825 "

Total cast-iron **15,549** "

Malleable Iron.

104 diagonal ties, each 9·2 ft. long, cross-section $\frac{2\,5}{1\,6}$ 4,500 lbs.

10 lateral rods each **17·7 long by 1 inch diam.**   234 lbs.
104 small **bolts for top chord each 6 inches long by**
    ¾ **inch diam.**   490 "
104 small bolts for bottom chord each 4 inches long
    by ¾ inch diam.   326 "
218 nuts **each** ½ lb.   109 "
264 lineal feet bottom chord **4 × ½**   1,584 "
                                                        ―――――
                                                        7,243

Wood.

12 floor-beams 7 × 14 14 feet long   1,372 ft. B. M.
2 track-strings 8 × 10 **66**   "   880   "
1800 feet B. M. floor plank   1,800   "
                                         ―――――
                Total board measure   4,052

Approximate Estimate of Cost.

15,549 lbs. castings @ 2½ cts. $388 72
7,243 "   rolled iron @ 4 cts.   289 72
4,000 feet B. M. lumber @ $15   60 00
Making **eyes on 104 diagonal ties** @ 30 cts. 312 00
   "   **10 lateral bolts** @ 25 cts.   2 50
   "   **208 small**   "   @ 10 "   20 00
Workmanship in fitting and raising **66** feet @ $5 330 00
                                              ―――――――
                                             $1,403 74

Estimated cost per foot lineal $21 75.

Calculation.

The weight of the bridge as determined by
    the bills of materials is   35,000 lbs.
Weight of maximum load **1 ton per foot** 132,000 "
Total weight   167,000 "
Weight on one-half the bridge   83,500 "
The pressure on the upper chord is   192,500 "

And is equivalent to 12,000 lbs. sq. in.
Tension on lower chord 48,000 "

The ties and braces form three distinct systems. The proportions of weight sustained by each may not be equal, and cannot be estimated with certainty, as one system may be brought into a higher degree of tension than another by driving the wedges unequally. In making a calculation, however, it will be assumed that they bear equally and each one-third of the weight.

The weight at the end being 83,500 lbs., the tension in the direction of the diagonals will be 116,900 lbs., or 38,966 lbs. to each system. This is resisted by two ties, the united cross-section of which is three inches, making the tension 12,988 lbs. per square inch.

The result of this calculation shows, that with the dimensions assumed the ties are stronger than the chords, and that heavier proportions are required to sustain a load of one ton per foot in addition to the weight of the structure.

For lighter loads the bridge is sufficient, and by increasing the dimensions, the trusses can be made as strong as may be necessary for ordinary spans.

When the top chord extends above the roadway so that it cannot be braced laterally, it is very important that its horizontal dimension should be increased as much as possible to prevent lateral flexure.

The dimensions used in the calculation are those of a bridge at 109th street in the city of New-York, as reported to the writer; they may not be entirely correct in every particular. The calculation has been made for a bridge of two trusses, for the sake of uniformity, as the other calculations have been made in the same way.

CUMBERLAND VALLEY RAILROAD BRIDGE

ACROSS THE RIVER SUSQUEHANNA, AT HARRISBURG.

The original contract price for the erection of the superstructure of the bridge was $52,000; but, in consequence of various accidents, the actual cost of construction was increased to $62,000.

It was used, from the time of its completion until Dec. 4, 1844, for locomotive engines, and was without roof; on this day a fire occurred, which destroyed all but four spans on the Harrisburg side of the river. It was quickly rebuilt on the same general plan as the original structure, with some slight alterations in the details; the hand-rail was omitted on the top, and a pointed roof substituted. On the new bridge locomotives are not allowed to pass.

As at present constructed, the bridge is an ordinary double lattice, the spans vary in length from 170 to 180 feet. There are 23 spans in all, and the total length of the bridge is 4,277 feet, making the average 186 feet from centre to centre of pieces, or 176 feet in the clear. The bridge is graded with one inclination towards the eastern shore, of 19 feet 10 inches in length of the bridge. There are two roadways on the lower chords, each 11 feet 1 inch from centre to centre of trusses, or 9 feet 1 inch in clear of chords.

Between the carriage-ways was a space of $6\frac{1}{2}$ feet, designed for the accommodation of foot-passengers, but it was found necessary to use this space for diagonal bracing. A single railroad track is on the top of the bridge, supported by the middle trusses, which are double lattice, while the outside trusses are single.

The trenails, or lattice-pins, are of oak, $1\frac{1}{2}$ inches in diameter, there are 4 at each intersection of the chords, and 3 at the intermediate intersections.

The total height of the outside trusses, from the top of the upper chord to the bottom of lower chord, is $18\frac{1}{2}$ feet. From the middle of the upper chord to the middle of the second chord is 2 feet 9 inches.

Total length of middle trusses 15 feet 9 inches.

Bill of Timber for One Span of 186 feet.

13,392 lineal feet chord plank	3	×	12		40,176 ft
176 lattice plank for outside trusses	3	×	9	24 ft. long	9,504 "
352 lattice plank for inside trusses	3	×	9	20 "	15,840 "
88 lower floor-beams	4	×	10	11 "	3,227 "
22 " foot-path	4	×	10	6 "	440 "
44 upper floor-beams	5	×	7	20 "	2,569 "
372 lineal feet track strings	6	×	8		1,488 "
22 lower cross-pieces between middle trusses	5	×	7	8 "	513 "
22 upper cross-pieces between middle trusses	5	×	7	6 "	385 "
88 knee-braces	4	×	5	5 "	739 "
12 diagonal braces	5	×	6	$12\frac{1}{2}$ "	375 "
12 "	5	×	6	10 "	300 "
44 roof braces	4	×	3	7 "	308 "
44 "	4	×	3	9 "	396 "
88 rafters	4	×	5	17 "	2,500 "
10,000 lineal feet lath	$2\frac{1}{2}$	×	1	"	2,083 "
88 lower lateral braces	2	×	6	10 "	880 "
88 upper "	3	×	7	$10\frac{1}{2}$ "	1,617 "
8,500 feet B. M. $2\frac{1}{2}$ inch oak floor plank					8,500 "
5,500 " " 1 " boards for upper floor					5,500 "
5,500 " " 1 " " sides					5,500 "
2 wall plates	5	×	12	30 ft. long	300 "
10 bolsters	7	×	9	15 "	789 "
2 pier pieces	4	×	10	24 "	160 "
					104,089 "

Also 26,000 shingles

1,760 pins, 18 inches long, $1\frac{1}{2}$ diameter
1,056 " 30 " $1\frac{1}{2}$ "
1,320 " 6 " $1\frac{1}{2}$ "

Iron Rods.

12 brace rods	1 in. 13½ ft. long	162 ft.	⎫
12 "	1 " 7¼ "	87 "	⎪
12 upper floor rods	1 " 20 "	240 "	⎬ 736 lineal ft.
12 " "	1 " 13⅓ "	160 "	⎪
12 " "	1 " 7¼ "	87 "	⎭

Total weight of iron   1,950 pounds.
"   timber (one span)   312,267 "
"   shingles   36,720 "

Total weight of one span   350,937 "
Weight per lineal foot   1,900 "

Estimate of Cost of One Span.

104,089 feet B. M. timber	@ $10½		$1092 93
26,000 shingles	@ 10½		273 00
1,950 pounds iron rods	@	4 cts.	78 00
Work on 60 rods and nuts	@	57 "	30 00
Workmanship on 186 lineal feet	@	6 58 "	1224 00
			$2697 93

Average cost per lineal foot $14 50.
Cost per lineal foot of single track bridge $8 00.

Data for Calculation.

As the middle trusses sustain the weight of the railroad track, and also one-half of each of the carriage-ways, they will bear a greater proportion of the load than the trusses on the outside, and, therefore, the calculation will be made for them. The data for calculation in this case will be,

Span between supports 176 feet.
Cross-section of upper chords, each 216 inches.
       "   lower "   "   108   "

From centre to centre of top and bottom chords 14 ft. 9 in
"   "   "   middle   "   9 " 3 "
Number of intersections in one span between supports, 42.
Proportion of weight of bridge on middle trusses, 175,000 pounds.

Greatest accidental load from a train of cars and two loaded wagons in middle of span, 200,000 pounds.

Total weight on one span—two trusses, 375,000 pounds.

Distance of centre of gravity from end, 45 feet.

The spans being framed continuously, the resistance of the chords at the piers may with propriety be taken into consideration, and the effect will be equivalent to doubling the resisting areas of the chords in the centre.

It will, therefore, be assumed that the resisting area of each of the upper and lower chords will be 648 square inches.

The neutral axis in this case will be in the centre of the trusses. If the strain per square inch at the extreme chords be represented by P, the strain on the middle chords will be $P \times \frac{4 \cdot 6}{7 \cdot 4}$: and the equation of equilibrium will be $648\ P \times 7\cdot 4 \times 2 + 403\ P \times 4\cdot 6 \times 2 = 187{,}500 \times 45$; or $13{,}297\ P = 8{,}437{,}500$; or $P \times 635$ pounds = strain per square inch upon the chords.

*Strain upon the **Ties**.*

The strain upon the diagonal ties and braces is very difficult to estimate correctly. The following considerations will, perhaps, lead to nearly correct conclusions in reference to the principle upon which a calculation may be attempted.

1. Whatever may be the particular arrangement of the parts, if the weight is uniformly distributed, there must be a gradual increase of vertical pressure from the middle to the ends.

2. By reference to the plate it will be perceived that there are six separate and independent systems of ties and braces, each similar to that exhibited in the annexed figure.

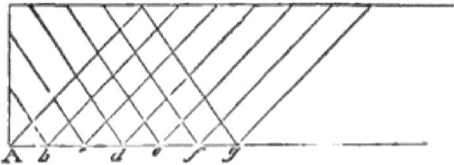

3. These systems, even if the workmanship be supposed to be perfect, cannot assist equally in sustaining the load, but the portion sustained by each will be nearly as the weights upon the points b, c, d, e, f, and g. In the present case, there are 21 spaces between A and the middle of the span, and the weight at g will be $\frac{15}{21}$ of the weight at A, and as the whole weight must be sustained by the systems which terminate between A and g, the portion upon one system at A must be 21 $\div (16 + 17 + 18 + 19 + 20 + 21) = \frac{21}{111}$ of the weight of one-half the span, or nearly one-fifth the weight.

The strain upon the diagonal is 1·4 times the vertical pressure, therefore, $\frac{1}{5} \times 1\cdot 4 = \frac{14}{50} =$ greatest proportion of weight sustained by any one system.

As the weight of the half span loaded, is 187,500 pounds, we will have $187{,}500 \times \frac{14}{50} = 52{,}500$ pounds.

As there are four truss frames, each will bear one-fourth, and $\frac{52{,}500}{4} = 13{,}125$ pounds is the greatest force either of tension or compression that any single lattice plank will be required to sustain.

The resisting cross-section of each plank, after deducting pin-holes, is 15 square inches, and the strain per square inch will consequently be 875 pounds.

The lattice plank in the direction of one of the sets of diagonals being in a state of compression, and the others in a state of tension, the effect upon the truss is to produce torsion. And it is generally observed that ordinary lattice bridges yield by twisting or warping before they fail in any other way.

16

TRENTON BRIDGE. (*Plate* 9.)

This bridge was built across the Delaware River at Trenton in the year 1804, by Lewis Wernwag. It is supported by five trusses, leaving four intervals for two carriage-ways and two footpaths. Each truss consists of a single arch composed of eight **planks 4 × 12 placed** in contact with each other. The roadway is suspended by chains of $1\frac{1}{8}$ inch square iron, the links of which are about 4 feet long, and 5 inches wide, passing flatways through the arches and between the chords and counter-braces; a key passing through the link on the top of the arch.

The counter-braces are in pairs 6 × 10, spiked to the chords at the lower ends, and connected with the arch at the upper end by means of iron straps $2 \times \frac{1}{2}$ inch.

The chords are also in pairs $6\frac{1}{2} \times 13\frac{1}{2}$, placed in contact; between them the links of the suspension-chains pass.

The floor-beams are suspended below the roadway under the cnords, and held in place by the suspension-chains, the lower links of which pass around them.

The chords are connected with the arches at the end, by means of long straps of iron passing around the end of the arch at the skew-back, and bolted through the chords. The width of each carriage-way in the clear is 11 feet, and of each footpath 6 feet.

On the sides are large **spur arches, of** the same dimensions as the main arch of the truss, extending from a point 8 feet outside of the truss on the abutments and piers, and terminating within 44 **feet of** the centre—spiked at the point of intersection to the arches of the main **truss.** During the present **year** changes have been made by removing the outside truss on the lower side, to a sufficient distance to convert the footpath into a carriage-way. Cast-iron shoes were also placed under the ends of the counter-braces.

The bridge is without lateral braces; its width appears to be sufficient to prevent lateral motion. **The** spur arches also assist in resisting the force of **the wind.**

DESCRIPTION OF AN IRON ARCHED BRIDGE OF 133 FEET SPAN,

ACROSS THE CANAL ON SECTION FIVE OF THE PENNSYLVANIA CENTRAL RAILROAD.

The chief peculiarity of this **bridge consists in its** *iron arch*, which is extended to a very considerable span, and furnishes a highly important practical test of the powers of resistance, both of the material itself and of the particular form **in** which it is employed. At the same **time,** the application of the principle upon which the structure is **erected** has been made under circumstances which render it perfectly safe; for in the event of the failure of the arch, the truss, without it, **is** more than sufficient to sustain the **greatest** load that **can** come upon the bridge.

The general arrangement **of** the truss is that of a Howe bridge, consisting of **top and** bottom chords of wood, with braces, counter-braces, and vertical rods. The braces are in pairs, and the arches pass between them. The counter-braces rest upon the arches, and are adjusted by means of set screws above and below.

The arch **is constructed of a centre rib** of cast-iron, 7 inches deep, **with** upper and lower horizontal flanches, 5 inches wide; **two** rolled iron plates are placed on the top, and two on the bottom of the cast rib, breaking joint with the rib and with each other, and secured by clamps at proper intervals. Below the chords are solid cast-iron skew-backs; and castings, of suitable **form** to connect with the skew-back and receive the ends of the arch, **are** placed on the top of the ower chord.

Believing that **the failure** of cast-iron bridges results gene rally from the inequality of pressure upon the joints, it was proposed to obviate this difficulty in the present case, by interposing plates of annealed copper between the ends of the segments, so that if the arch should rise or fall by expansion or contraction, the comparatively yielding quality of the interposed material would distribute the pressure, and prevent the fracture which might be produced if the joint should open, and the pressure be thrown upon the upper or lower corners of the castings.

This intention was defeated by circumstances which rendered it necessary to hasten the completion of the work. The ribs were raised without dressing the joints, and the copper plates were therefore rendered useless, the inequalities of surface being too great to admit of their being advantageously employed. Under these circumstances a substitute was used, which gave more satisfaction than could have been obtained by an adherence to the original design, and was much more economical. The joints were separated to the distance of one-fourth of an inch, and filled with spelter poured into them in a melted state; this was very conveniently done by binding a piece of sheet-iron around each joint, and covering it with clay. The material introduced being nearly as hard as the iron itself, and filling all the inequalities of the surface, rendered the connection perfect.

The pieces of castings were made with inch holes near the ends, through which rods were passed horizontally to assist in raising them. To support them when raised to their proper positions, pieces of board were nailed vertically from the top to the bottom chord, on each side of the truss, and short rods were passed through the holes in the ends of the castings, and through augur holes in the boards. By this arrangement the segments were held securely, and no obstruction was offered to the attachment of the arch-plates, which were added by clamping one end, and springing them around the arch by a rope attached to the other.

The most important advantage that was expected to be derived from the peculiar arrangement exhibited in this structure, was a practical test of the power of resistance of a counter-braced iron arch on a large scale, under circumstances which would render its failure, under any possible contingency, unattended with risk.

It has been observed that counter-braces are placed above the arch, resting against it by means of adjusting or set screws. In addition to this, there is a vertical post of oak between each pair of suspension rods, also terminating in a set screw resting on the arch. It will be readily perceived that, by loosening the lower counter-brace screws, and by tightening those on the posts, the bridge will be raised upon the arch, and the latter will then bear the whole weight both of the truss and its load. If the arch should prove unable to sustain this pressure, the truss would sink again to its original position and receive the weight.

The experiment thus far has been entirely successful, and shows that the counter-braced arch, which is the lightest and cheapest system possible, is perfectly reliable for spans of any magnitude; it is, in fact, a satisfactory test both of the principle and of the material.

The manner of adjusting the trusses, and the observations subsequently made, were as follows:—

1. All the lower counter-braces were unscrewed.

2. All the upper counter-braces were tightened, but not screwed hard.

3. The levels were taken from a permanent level mark at the foot of every suspension rod.

4. The set screws upon the posts were tightened by two men, with suitable wrenches, beginning at the middle and proceeding towards the ends; after once going over, the level was again taken, and the bridge found to be raised one-fourth of an inch. The same operation was repeated, and the rise found to be half an inch, which was sufficient to make it certain that the whole weight was upon the arch.

5. The upper counter-braces were examined and found to have become loosened; they were again screwed up; the main-braces were loose;—all of which were necessary consequences of raising the truss upon the arch. The time at which these adjustments were made was about 11 o'clock, in July; the weather warm.

6. The bridge was again examined on **the following morning**; the weather was cool, **and the** contraction of the arch, from difference of temperature, had caused the posts and counter-braces to **become loose**, whilst the main-braces were found to be in full action; in other words, by the contraction of the arch the weight was again thrown upon the truss.

7. The post and upper counter-brace screws were once more tightened, and as the heat of the day increased, the arch expanded and lifted the bridge to a greater height than on the preceding day.

8. While in this condition, the whole weight being upon the arch, the posts and upper counter-brace screws tight, and the lower counter-brace screws loose, a locomotive was passed over **the bridge, and** observations made with a level and **rod whilst it was running** repeatedly backwards and forwards. The greatest variation from the level of repose was $\frac{1}{8}$ inch. The arch rose slightly when the locomotive was upon the opposite side, and fell as much below its **original position when it was on** the **same side as** that upon which **the observation was made**. This was the effect anticipated; it **was** not to be supposed that the arch, in the condition it then was, would be perfectly rigid, as the lower counter-braces were all unscrewed, and the upper ones, as their resistance was not transmitted to the opposite extremities of the diagonals of the panels, could not act with full effect.

9. After an interval of several days, during which the arch **and** truss experienced no change, except that the lower counter-braces were screwed up, a 23-ton locomotive was passed several times over the bridge. No level was at hand with which to make observations instrumentally, but the eye could detect no motion in the arch; it appeared to be perfectly rigid in every direction, and a subsequent careful examination after more than one year of service cannot detect the slightest opening or compression at the joints.

The observations made thus far have been sufficient to satisfy the writer of the correctness of his views in regard to the strength and rigidity of a counter-braced arch, and its applicability **to spans** of great extent. Upon this principle, an iron bridge can be constructed **at less expense than is** now sometimes incurred in the erection of wooden ones, and the durability, with proper care, is almost unlimited.

An iron arch, constructed in a manner similar to the above, would be perhaps the cheapest and best support for an aqueduct. As the load in this case is always nearly constant and uniform, the curve of the arch should be a parabola.

No practical difficulty need result from expansion and **contraction**, particularly if iron tie-rods are not used for the lower chords. The counter-brace rods can be so proportioned and disposed as to compensate **for** changes in the arch, **and keep** the tension constant.

IRON BRIDGE OVER RACOON CREEK,

PENNSYLVANIA RAILROAD. (*Plate* 11.)

This bridge depends for **its** support **upon 4 counter-braced** arches, constituting a single system, unconnected with **any self-supporting truss.** The arches are in pairs, one on each side of each truss; they are composed of plates of malleable iron, **1** inch by 3 inches, placed one upon another—3 at top and 3 at the bottom of each, separated by pedestals and diagonal braces, and secured in place by wrought-iron clamps and **bolts.** There are, consequently, in the two trusses 24 leaves **or plates, 1 by** 3, arranged in **groups of 3, the separate plates** breaking joints with each other.

The diagonal-braces **between the arches are connected by** iron keys, kept **in place by the clamp bolts.**

Each skew-back **has 4 box-shaped cavities to receive the ends of** the **plates.**

The **top chord is of wood, 12 by 12** inches. The lateral, diagonal, and counter-braced rods pass through it, and are so cured by cast-iron angle-blocks, and nuts on the outside.

The roadway **is** on **top.** The weight is transmitted to the **arch by** means **of** hollow columns or cylinders. Each cylinder is capped with a circular plate, which is cast with projections on both sides, fitting into the bottom of the chord and into the top of the column. The lower ends of the columns rest in sockets. The socket-boxes have cylindrical projections, 3 inches in diameter, upon the sides, which fit into openings in the pedestals, and the pedestals rest between the arches, being firmly held in place by the clamps and diagonal braces. The cylindrical form of the socket-box admits of a vertical position for the post at every point.

This description cannot readily be understood except by reference to the plates.

The small posts or columns which connect the system of counter-braces pass entirely through the socket-boxes and outside cylinders, and are of uniform length, extending from the top to the bottom chord. They are connected with the bottom chord by the ends of the diagonal rods, which pass through them and serve as bolts.

The action of the system is the reverse of that which takes place in an ordinary truss. There is no tension on the lower chord in the middle; this may be disconnected without injury. A strain upon the counter-brace rods produces a tension on the lower chord at the skew-back, with which it is securely connected, but none in the middle of the span, where there is a coupling link to allow of expansion and contraction.

The lower lateral bracing is by means of diagonal rods and cross-beams of iron; the upper bracing consists of wooden braces and lateral rods perpendicular to the chords.

Bill of Materials.

CAST-IRON.

4 skew-backs	592 pounds
28 pedestals	714 "
32 right diagonal arch braces } 32 left " " }	2,720 "
2 exterior cylinders $3\frac{3}{16}$ inches long, diameter $3\frac{3}{4}$ and $5\frac{3}{4}$ inches	22 "
Amount carried up	4,048 "

	Amount brought up	4,048 pounds.
4 exterior cylinders 6¾ inches long, diameter 3¾ and 5¾ inches		96 "
4 exterior cylinders 18 1/16 inches long, diameter 3¾ and 5¾ inches		264 "
4 exterior cylinders 36 13/16 inches long, diameter 3¾ and 5¾ inches		556 "
14 interior cylinders, 5 ft. 6 11/16 in. long, diameter 3½ inches		1,596 "
14 cap-plates		98 "
14 socket-boxes		476 "
7 girders		805 "
14 lateral-brace blocks		126 "
34 small angle blocks		69 "
34 keys for arch-braces		85 "
	Total weight of castings	8,219 "

MALLEABLE IRON.

62 plates for arches, 3 × 1, various lengths to break joint 11,559 pounds.		
86 coupling plates, ¾ × 2¼, 16 inches long		655 "
192 ¾ in. bolts with head and screw, 14½ in. long		358 "
192 nuts 2¼ × ¾		305 "
4 skew-back bolts bent at an angle of 45° at one end, and furnished with nut and screw at the other end, the bent end having an eye to receive the end of the lateral brace rod, 17 inches long		63 "
4 coupling links for lower chords		33 "
4 rods for horizontal bracing, 1st panels 3 inches at one end, bent at an angle of 90° to pass through the skew-back bolt; nuts and screws on both ends, length from angle 7 ft. 11½ inches.		
	Amount carried over	1,314 "

Amount brought over 1,314 pounds

4 lateral-brace rods for 2d panels, 8 feet 11 inches.

4 lateral-brace rods for 3d panels, 9 feet long.

4 " 4th " 9 feet ½ in. long.

14 diagonal-brace rods with nut and screw at upper end, lower end bent at an angle of 49° from straight direction, to pass through the column and tie, secured by nut and screw on outside, length of whole rod 9 feet 11 inches, elbow 7 inches.

4 counter-brace rods for 1st panels, the upper end passes through angle-block placed on top of chord with nut and screw; the lower end is formed into an eye to embrace the skew-back bolt, length from centre of eye, 8 feet 11 inches.

4 counter-brace rods for 2d panels, upper end as before; lower end bent 138° to enter the bottom of the cast-iron column, the bent end has an eye 2 inches long, 1⅛ inch wide, the centre of which is 2¼ inches from upper side of angle of rod, length of rod from angular point to end of screw 9 feet 4 inches. Total length 9 feet 8 inches.

4 counter-brace rods for 3d panels similar to those in 2d panels, length 9 feet 5 inches.

4 counter-brace rods for 4th or middle panels as above, length 9 feet 5½ inches.

7 rods for upper lateral braces, 7 feet 3 inches.

Amount carried up 1,314 "

Amount brought up	1,314 pounds.
Total length of the 5/8 inch rods, 483 feet, weight	1,280 "
4 lower chords 48 feet long, 1¼ inches diameter,	705 "
100 nuts for inch bolts 1 × 2¼ × 2¼	141 "
Total weight of malleable iron, exclusive of arch plates	3,440 "

TIMBER.

2 upper caps 12 × 12, 50 ft. long	600 feet B. M.
16 lateral-braces 4 × 5, 8½ "	126 " "
	726 " "

Estimate.

8,219 pounds castings at 2 cents	$164 38
11,559 " arch plates at $57 per ton gross	294 11
3,440 " bolt and plate iron at 3½ cents	120 50
726 feet B. M. timber at $12 per M.	8 71
Total cost of materials	$587 70

Workmanship.

69 days' work making plates and fitting pieces at $1 50	$103 50
Files	3 50
Making 5/8 inch bolts at 53 cents	13 25
" 192 ¾ inch bolts at 10 cents	19 20
Freight and tolls for delivery of materials	21 00
4 men 6 days raising at $1 25	30 00
Total cost of workmanship	190 45
Cost of materials per foot of span	12 50
" work "	6 05
Total cost per foot	$18 55

Data for Calculation.

Span   47 feet.
Rise of arch   5 "
Cross-section of all the arches in square inches, exclusive of castings   72 "
Distance of centre of gravity from abutment   12 "
Whole weight of bridge   27,500 pounds.
Weight of bridge and load   122,000 "

As the arch sustains the whole of the weight, the calculation is extremely simple.

Let $P =$ pressure per square inch at crown.

Then $P \times 5 \times 72 = \dfrac{122000}{2} + 12$, or $P = 2033$ pounds per square inch, or only about one-thirtieth of the crushing force.

The greatest pressure upon any one post may be taken at 6 tons. The cross-section is 15 square inches. Pressure per square inch 800 pounds.

The projections of socket-boxes are 3 inches cylinder, the pressure on each is 3 tons = per square inch 630 pounds.

It is unnecessary to calculate the strain upon the counter-brace rods, they are evidently sufficient; and for the manner of making the calculations, sufficient illustrations have already been given. As a general rule in regard to counter-braces, it may be stated, that their dimensions may be assumed as constant, whatever may be the span; or rather, the counter-braces should bear a fixed proportion to the width of the panels, without reference to any of the other dimensions of the bridge, and, consequently, the counter-brace rods need not be larger or more numerous in proportion to the length in a bridge of large span than a shorter one.

The truth of this assertion will be evident from these considerations:—

The greatest possible strain upon any counter-brace has been shown to be less than the variable load upon one panel. The weight of the structure produces no strain whatever upon the counter-braces. The greatest variable load on railroad bridges has been assumed as one ton per foot lineal, which will be one thousand pounds per foot lineal for each truss. If the panels are 10 feet (which is nearly an average for the bridges on the Pennsylvania Railroad), the greatest strain upon any single counter-brace will be 10,000 pounds, and this will be resisted by a single square inch of metal.

As a general rule, which will save much trouble in calculation, the proper cross-section of the counter-brace rods for railroad bridges of any span may be estimated at one square inch for every 10 feet of truss.

BALTIMORE AND OHIO R. R. BRIDGE. (Plate 12.)

The plan of this bridge was furnished by B. H. Latrobe, Esq., Chief Engineer. It is an admirable combination, possessing every essential of a well-proportioned and scientifically arranged structure. It is a system of counter-braces and braces. In its general principle it bears some resemblance to the celebrated bridge across the Rhine at Schauffhausen, but the latter, owing to the absence of counter-braces, was so flexible that it would vibrate with the weight of a single man, whilst the Baltimore and Ohio R. R. Bridge is so rigid that the heaviest locomotives, running with great velocity, produce but very little effect.

These bridges possess great strength, but they are not as economical in first cost as many others.

The calculation for the strains is more simple than in any other form of bridge; each set of arch-braces is to be considered as sustaining one-half the weight of the interval on each side of it, between it and the next set of braces.

Description of Details.

Fig. 1 shows the manner of adjusting the horizontal diagonal brace, in tie-beams.

Fig. 2 shows the manner of adjusting the horizontal braces in floor-beams.

Fig. 3 represents the skew-back, which is cast in two pieces, the hindmost part $a\,a$ being the buttress of the main part $b\,b$. The heels of the arch-braces rest in cast-iron shoes, between which and the abutting steps of the skew-back, adjusting-screws operate to push the braces forward when required in raising or adjusting the truss.

Fig. 4 shows the arrangement of the chord-splices.

Fig. 5 **shows the** intersection of the counter-brace and main-braces. **The counter-braces are cut off at** their intersections **with the main or panel-braces, and** their connection carried **around the latter by means of the cast plates** shown at $d\,d$, these plates are **connected across the truss by bolts pass**ing within cast tubes acting as struts, as at $f\,f$.

Fig. 6 shows the connection of the upper tie-beams with the chords, braces, and counter-braces.

The principal rafters foot upon the casting c between the tie-beams.

Fig. 7 **shows** the manner **of** securing **the rail to the** rail-joist or string-piece. The rail-joists and floor-beams **are** tied together by a vertical bolt at each intersection. **The** rail is fastened to the rail-joist by double-headed bolts.

All **the** principal abutting surfaces of the timbers are separated **by cast-iron** plates, and **every** joist has an independent **adjustment by means of screw-bolts** or wedges.

Material in one span of 133 feet in the clear of abutments, or 145 feet from end to end **of skew-backs:**

Timber, 63,000 feet B. M.
Cast-iron, **57,156** pounds.
Wrought-iron, 15,340.

CANAL BRIDGE, SECTION 6, PENN. RAILROAD.

This bridge has some resemblance to that on Sec. 5, exhibited in Plate **10**. **The principal** differences are the absence of posts, and the use of a wooden arch composed of layers of plank, instead of an arch of cast-iron.

The advantage of such an arrangement is the great facility which it affords for adjustment. To raise the camber of this bridge it is not necessary to remove a single stick of timber; all that is required is to slacken the counter-braces and tighten the vertical rods, until the bridge is raised to a sufficient extent, after which the counter-braces should be tightened. Two men in less than an hour can adjust a bridge constructed in this way.

The arch can be made to bear any proportion of the weight by tightening the counter-braces on the upper side.

BOILER PLATE TUBULAR BRIDGE. (*Plate* 4.)

(COPY OF A LETTER FROM THE INVENTOR.)

Reading, May 1, 1849.

DEAR SIR:—Inclosed I send you the drawings of the three bridges I constructed on the Baltimore and Susquehanna Railroad while engaged as Superintendent of Machinery and Road.

The one marked A was built at the Bolton depot in the winter of 1846 and '7, and was put in its place in April, 1847. This bridge is made of puddled boiler-iron $\frac{1}{4}$ inch in thickness. The sheets, standing vertical, are 38 inches wide and 6 feet high, and riveted together with $\frac{5}{8}$ rivets, two and a half inches from centre to centre of rivets. You will observe by reference to the drawing, that each truss-frame is composed of two thicknesses of iron, 12 inches distant from each other, and connected together by $\frac{5}{16}$ iron bolts, passing through round cast iron sockets at intervals of 12 inches; which arrangement, together with the lateral bracing between the two trusses, which is composed of $\frac{3}{4}$ round iron, set diagonally and bound together at the crossing by two cast-iron plates about 4 inches diameter, the sides next to the bracing being cut in such a manner, that when the two ⅜ bolts that pass through them were screwed up, it held them firmly together. There is also a bolt passing through both truss-frames and through the heels of the lateral bracing, at right angles with the bridge, which secured the heels of the lateral braces, and by means of a socket in the centre made a lateral tie to the bridge, giving the bridge its lateral stability. The lower chords were of hammered iron, there being some difficulty at that time to get rolled iron of the proper size, and are in one entire piece, being welded together from bars 12 feet long. There are eight of them 5 × ¾ inches, one on either side of each piece of boiler iron, and fastened to it with ⅜ inch iron rivets 6 inches distant from each other. There are but four top chords, and of the same size of the bottom, two on each truss near the top, the timber for the rail making up the deficiency for compression, and answering the purpose of chords. This bridge was built at the time Messrs. Stephenson and Brunell were making their experiments with cylindrical tubes preparatory to constructing the Menai bridge; the cylindrical tubes failing, they adopted this plan of bridge. The entire weight of the bridge is 14 gross tons, and cost $2,200; but as the same kind of iron of which the bridge is composed can be had for at least 15 per cent. less now, than it cost at that time, it would be but fair to estimate the cost of the bridge at $1,870, without any reference to the labor that is misapplied in all new structures of the kind, making the cost of a bridge 55 feet long $34 per foot. And I have no doubt, where there would be a large quantity of iron required for such purposes, that it could be had at such prices as to bring down the cost of bridges of 55 feet length to $30 per foot.

Very respectfully yours,
JAMES MILLHOLLAND

ARCHED TRUSS BRIDGE, READING RAILROAD.

(*Plate* 2.)

(DIRECTIONS GIVEN BY PATENTEE, J. D. STEELE.)

This improvement consists in combining arches with a truss frame by securing them to tension posts—"*a a*," which posts are connected to the chords by screw-fastenings, "*e e*," and so arranged as to admit of changing the position of the arches relatively to the chords, or of drawing together the chords without changing the position of the arches, by which means the strain can be regulated and distributed over the different parts of the bridge at pleasure.

In erecting a bridge on this plan it will be found desirable to be governed by the following directions, viz. :— The truss must first be erected, provided with suitable cast-iron skew-backs to receive the braces and tension posts, and the several parts of the chords should be connected with cast-iron gibs. Wedging under the counter-braces must be avoided by extending the distance between the top skew-backs sufficiently to bring the tension posts on the radii of the curve of cambre of the bridge. The tension posts must be about eight inches shorter than the distance between the chords, and in screwing up the truss care must be taken not to bring their ends in contact with the chords; but they must be equidistant, and about four inches from them. When the truss is thus finished it must be thrown on its final bearings, and it is then ready to receive the arches, which should be constructed on the curve of the parabola, with the ordinates so calculated as to be measured along the central line of the tension posts. They must be firmly fastened to the posts and bottom chords by means of strong screw-bolts and connecting plates, as shown at "*d d*," and should foot on the masonry some distance below the truss, which can be done with safety, as the attachment to the posts and chords will relieve the masonry of much of their horizontal thrust. When a bridge so constructed is put into

17

use it will be found, as the timber becomes seasoned, the weight will be gradually thrown upon the arches, which will ultimately bear an undue portion of the load. To avoid this the cambre must be restored and the posts moved up, so as again **to divide** the strain between the truss and the arches.

This adjustment must take place once or twice **in each year, until** the timber becomes perfectly seasoned, after **which,** in a well constructed bridge, but little attention will be required. Plates of iron should in all cases be introduced **between the** abutting surfaces of the top chords and arches, and all possible care taken to prevent two pieces of timber from coming in contact, by which decay is hastened; care should also be taken to obtain the curve of the parabola for the arches, **as it is** the curve of equilibrium and of greatest strength, as has been shown by experiment.*

Bridges constructed on this plan will be found to possess **an** unusual amount of strength, for the quantity of material **contained in them, and if well** built and protected, great durability.

BRIDGE ACROSS THE SUSQUEHANNA, AT CLARK'S FERRY (*Plate* 13.)

This bridge **has** been selected as a fair specimen of the ordinary Burr bridge, a mode of construction more common than any other in Pennsylvania, and which experience for many years has proved to be **one of the** best arrangements for ordinary purposes.

Probably no other plan has ever secured more general approbation or better sustained itself than the Burr bridge, and

---

* **The** parabola is the curve of equilibrium when no load is upon the bridge, and also when the load is uniform, but there can be no curve of equilibrium for a variable load of a passing train. Stiffness can be secured in this case, only by an efficient system of counter-braces.

The plan proposed fulfils every condition of a good bridge.—*Author* it is certain that when counter-braced and properly proportioned, it forms a truss fully sufficient to bear the heaviest railroad train. A particular description is unnecessary; the plates, with the accompanying bill of materials, will furnish the engineer or builder with all necessary information, and a sufficient number of examples have been given, to show how the calculations for the strains are to be made.

These bridges, well counter-braced, have been recently introduced into New England, by H. R. Campbell, Esq., and are remarkably rigid structures.

Bill of Timber for One Span.

		Inches.	Length. Feet. Inches	Feet.
2	Wall plates	10 × 15	28	700
30	Bottom chords	8 × 15	40 6	12,150
15	Top "	11 × 11	40 6	6,135
21	Floor beams (large)	10 × 11	36	6,930
21	" " (small)	5 × 11	36	3,465
36	Arch pieces	8 × 15	31	21,160
48	Queen posts	11 × 14	17 6	10,752
6	" "	11 × 14	23 6	1,812
6	" "	11 × 14	22	1,692
3	Ring "	11 × 18	17 6	867
60	Main braces	8 × 11	14	6,180
120	Check "	4½ × 11	9 6	4,800
21	Tie beams	9 × 10	28	4,410
84	Knee braces	4 × 5	8	1,125
80	Lower laterals	4 × 8	12	2,560
40	Upper "	4 × 8	24	2,560
120	Rafters 3 × 5 and	3 × 6	14 6	2,640
21	Roof posts	1½ × 11	3 6	100
6,912	Feet B. M. weather-boarding	3 × 12	13	6,912
22,260	Shingles			
9,540	Lineal feet laths			
	3 inch flooring plank		24	7,314
	Hand railing	2½ × 7		3,710

TOWING PATH.

	Inches.	Length. Feet. Inches	Feet.
7 Arch pieces	6 × 15	32	1,344
6 Lower chords	6 × 12	40 6	1,458
6 Upper " (railing caps)	5 × 10	40 6	1,014
1 Ring post	9 × 12	8	72
38 Queen posts	9 × 9	8	2,042
2 " "	9 × 9	8	180
2 " "	9 × 9	12 6	168
40 Braces	5 × 9	7	1,200
Weather-boarding, square feet			4,120
Flooring in lengths of		24	1,781

IMPROVED LATTICE BRIDGES. (*Plate* 14.)

Amongst the **great variety of** bridge plans that have from time to time been presented to the American public, none have experienced a more favorable reception than the ordinary lattice. Its beautiful simplicity, light appearance, and especially its economy, have secured for it the favor of many of our most **eminent** engineers and builders, as is evident from the extent **to which** it has **been** adopted on works **of** the greatest magni**tude and** importance.

On ordinary roads, and **on railways** not subjected to very **heavy** transportation, **this** plan of superstructure, when well constructed, has been found to possess almost every desideratum: nevertheless, **experience** has **fully** proved, that unless strengthened by additional arch-braces, **or** arches, the capacity of the **structure** is limited to light **loads,** and spans **of** small **extent. The** public works of Pennsylvania furnish abundant **proof of the** truth of this assertion ; and several railways might **be enumerated, on** which the lattice bridges have from neces**sity been strengthened by** props from the ground, by arches, **or arch-braces,** added when the insufficiency of the structure was found to require it. These circumstances have produced a change in opinion hostile to the whole plan, and it is much to be regretted, that instead of introducing such modifications and improvements as would remedy existing defects and retain its advantages, other plans have been substituted at an expense frequently more than double that of an efficient lattice structure.

One of the first defects apparent in some old lattice bridges, is the warped condition of the side trusses. The cause which produces this effect cannot perhaps be more simply explained than by comparing them to a thin and deep board placed edgeways on two supports, and loaded with a heavy weight; so long as a proper lateral support is furnished, the strength may be found sufficient, but when the lateral support is removed, the board twists and falls.

A lattice truss is composed of thin plank, and its construction is in every respect such as to render this illustration appropriate. Torsion is the direct effect of the action of any weight, however small, upon the single lattice.

A second defect may be found in the inclined position of the tie; all bridge-trusses, whatever may be their particular construction, are composed of three series of timbers; those which resist and transmit the vertical forces are called ties and braces, and those which resist the horizontal force are known by the names of chords, caps, &c.

In every plan except the common lattice, these ties are either vertical, or perpendicular to the lower chords or arches, and as the force transmitted by any brace is naturally resolved into two components, one in the direction of, and the other at right angles, to the chord or arch, it would seem that this latter force could be best resisted by a tie whose direction was also perpendicular. The short ties and braces at the extremities, furnishing but an insecure support, render these points, which require the greatest strength, weaker than all others; this defect is generally removed by extending the truss over the edge of the abutment a distance about equal to its height, or to such a distance that the short ties will not be required to sustain any portion of the weight, the effect of which is to provide a remedy at the expense of economy, by the introduction of from fifteen to thirty feet of additional truss.

A bridge whose corresponding timbers in all its parts are of the same size, is badly proportioned; some parts must be unnecessarily strong, or others too weak, and a useless profusion of material must be allowed, or the structure will be insufficient.

If, for example, the forces acting on the chords increase constantly from the ends to the centre, the most scientific mode of compensation would appear to be, to increase gradually the thickness of the chords; and for similar reasons the ties and braces should increase in an inverse order from the centre to the ends.

In accordance with this, it is found that in bridges that have settled to a considerable extent, the greatest deflection is always near the abutment; that is, the chords are bent more at this point than in the centre, and the joints of the braces are much more compressed. It is also found that **the weakest** point **of** a lattice bridge is near the centre of the lower chord; this might be expected, since from the nature of the force, and the mode of connection, the joints of the lower chords are only half as strong as the corresponding ones of the upper chord, it being assumed that the resistances to compression and extension are equal. This defect may be in a great degree removed by inserting wedges behind the ends of the lower chords. A variation in the size of every timber, according to the pressure **it is to** sustain, would of course be inconvenient and expensive; but as the principle of proportioning the parts to the forces acting upon them, is of great importance, such other arrangements should be adopted as will secure its advantages, **and** at the same time possess sufficient simplicity for practice, this is effected by the introduction of arch-braces or arches, than which, a more simple, scientific, and efficacious mode of strengthening a bridge could not perhaps be devised, as they not only serve, with the addition of straining beams, to relieve the chords, and give them that increase of thickness at the points of maximum pressure, which is essential to strength, but they also relieve the ties and braces by transmitting directly to the abutments, or other fixed supports, a great part of the weight that they would otherwise be required to sustain.

It may perhaps be objected, that the pressure of the arch-braces or arches would injure the abutments: in answer to this, it may be remarked that a certain degree of pressure is very proper; the embankment behind an abutment exerts a very great force upon it, the tendency of which is to push it forward; if, then, a counter-pressure can be produced by the thrust of arch-braces, or by wedging behind the ends of the lower chords, two important advantages are gained; the abutment is not only increased in stability, but the tension on the lower chord of the bridge is diminished by an amount equal to the degree of pressure thus produced.

It is, however, proper to observe, that when the situation of the embankment exposes it to the danger of being washed away from the back of an abutment, the pressure on its face must not be sufficient to destroy its equilibrium; should this effect be apprehended, the horizontal ties must be sufficient to sustain the thrust of the bridge.

An essential condition in every good bridge is, that it shall not only be sufficient to resist the greatest dead weight that it can ever be required to sustain in the ordinary course of service, but it must also be secure against the effects of variable loads. This is generally effected by the addition of counter-braces; but the lattice truss possesses this peculiarity, that it is counter-braced without the addition of pieces designed exclusively for this purpose: to prove this, invert the truss, when it will be apparent that the braces become ties, and the ties braces, possessing the same strength in both positions.

The foregoing remarks will, it is believed, enable the reader to understand the objects of the proposed improvements and the principles on which they are founded.

1st. The braces, instead of being single, as in the common lattice, are in pairs, one on each side of the truss, between which a vertical tie passes; this arrangement increases the stiffness upon the same principle that a hollow cylinder is more stiff than a solid one with the same quantity of material, and of the same length, and obviates the defect of warping.

2d. The tie **is vertical, or** perpendicular to the lower chord, a position which is more natural, and in **which** it is **more efficacious** than when inclined.

3d. The end-braces all rest on and radiate from the abutment, by which means a firm support is given to the structure, and the truss is not required of greater length than is sufficient to **give** the braces room.

4th. The truss is effectually counter-braced, the braces becoming ties, and the ties braces, when called into action by a variable load, and are capable of opposing a resistance on the principle of the inclined tie of the ordinary lattice bridge.

It is readily admitted that the strength in the inverted is less than in the erect position, but it must be remembered that **the unloaded** bridge is always in equilibrium; that the action of the parts which renders counter-bracing necessary, results entirely from the variable load, and that, therefore, **a combination** of timbers to resist its effects should not be as strong as that which sustains both the permanent and the variable loads.

Behind **the ends of the lower chords at the abutments, and** between them **over** the piers, double wedges are driven, the object of which is, by the compression which they produce, to relieve the tension of the lower chord.

For ordinary spans, the dimensions of the timbers may be:

    Braces   2 in. by 10 in. in pairs.
    Ties   3 " 12 "
    Arches or arch-braces 6 " 12 "
    Chords   3 " 14 " lapped.
    Pins $2\frac{1}{4}$ in. in diameter.

In conclusion, it is proper to remark **that the** proposed plan is not recommended as the best under all circumstances, but it is as economical in first cost as any other that can be used. The arrangement will be found even more simple than the ordinary lattice, and it is equally applicable for bridges on common roads or railroads, and for roof or deck bridges. The braces, in consequence of being placed in pairs, require a slight increase of timber over the common plan, in the proportion of 40 to 36, but the diminished lengths of the ties and of the truss **more** than counterbalance this increase.

The cost of workmanship on the truss is very trifling, and less than on the common lattice; if the timbers are cut to the proper lengths, the auger will be the only tool required in putting it together.

TRUSSED GIRDER BRIDGES. (*Plate* 15.)

Much diversity of opinion exists in regard to the manner of constructing trussed girder bridges, and the true relative proportion of the beams and tension rods. It is asserted by some that small tension rods are worse than useless, others think that even the slightest rod must render some assistance, and that the strength will be increased by any addition of this kind. A practical illustration of the subject can best be given by an example, and a calculation will be made from the following data.

Span between supports	50 feet.
Between points of attachment of rods	54 "
Middle interval	14 "
End intervals	18 "
Number of girders	2
Distance of middle of rod below middle of girder	3 feet.

The first hypothesis will be that the truss rods are two in number, each 1 inch in diameter.

The span being 50 feet, and the distance of the rod below middle of girder 3 feet, the length of rod between edges of abutments will be $14 + 2\sqrt{18^2 + 3^2} = 50,496$ feet.

The deflection of the beam itself, allowing the load to be 1 ton per foot uniformly distributed, and the weight of timber 3 pounds per foot B. M., will be determined from the expression $w = \dfrac{80\, b\, d^3}{l^2}$ in which the deflection is supposed to be $\frac{1}{4\text{ь}}$ of an inch for 1 foot in length, or $\frac{50}{48} = 1\frac{1}{4}$ inches in 50 feet; the weight being applied at centre. By substituting the proper values, we have $w = \dfrac{80 \times 20 \times 20^3}{50^2} = 5{,}120$ pounds.

The deflection of the beam, therefore, on the **supposition** that it does not break, but preserves its elasticity, **will be** 5,120 : 55,00 : : 1¼ : 13½ = total deflection in inches on this hypothesis.

But the **actual deflection of the trussed** beam will **not be as** great as this, since the rods will resist a portion of the strain, and consequently the beam will be to some extent relieved as long **as** the rods remain unbroken.

The result of experiments on the strength of wrought-iron is, that in perfect specimens the elasticity is not impaired, and consequently the strength is not injured by a weight which does not exceed 15,000 pounds per square inch. Of course a weight greater than the elastic limit must eventually cause the fracture of the material; *when it begins to lose its elasticity, it begins to break.*

Assuming the resistance therefore at 15,000 pounds per **square** inch acting as a force of tension on the rods, the weight which it would sustain at f or g', will be $\dfrac{15,000 \times 3}{18,248} = 2,470$ pounds, **equivalent** to a uniform load over the whole beam, of 17 : 50 : : 2,470 : : 7,265 pounds for every square inch in the cross-section of the **rods.**

The actual cross-section of the rods being 1·57 square inches, the **actual** weight that they can sustain will be 7,265 × 1·57 = 11,406 pounds. Deducting 11,406 pounds from 55,000 **pounds, there remain** 43,594 pounds to be sustained by the **beam itself.**

The deflection of the beam by this weight will be 5,120 : 43,594 : : 1¼ : : 10½ inches = the actual deflection of the trussed beam on the supposition that the rod will admit of a sufficient extension without breaking, or injury to its elasticity. This point must now be examined.

If the deflection becomes 10½ inches = $\tfrac{9}{10}$ foot, the length of the rod estimated as before, will be $14 \times 2 \sqrt{18^2 \times 3\cdot9^2} = $ 50,836 feet.

The extension will therefore be 50,836 − 50,496 = ·340 feet = $\tfrac{1}{147}$ **of the length of** the rod.

The extension that iron is capable **of** bearing without in jury to its elasticity is only $\frac{1}{1400}$ of its length. Consequently the strain with the dimensions assumed will be more than nine times the elastic limit. Therefore the rods must break before the beam reaches the deflection that the weight must necessarily produce, and are evidently of no assistance.

The truth of this conclusion is confirmed by experience, and practical men are adopting the opinion, based on experience and observation, that small rods, unless increased in number to secure a sufficient resisting area, are of no value.

The only correct and safe way of proportioning a trussed girder bridge is, to assume the size of the beams and all other dimensions of the structure except the rods, and determine the cross-section of the latter by the condition that the strain per square inch shall be a given quantity.

To illustrate this case let the same dimensions be continued, the rods excepted, and let the maximum strain per square inch be limited to 10,000 pounds.

The position of the neutral axis from which the strains should be estimated will depend,

1. On the relative magnitude of the cross-sections of the rods and girders.

2. On the relative powers of resistance of the material.

3. On the relative extensibility and compressibility of the material.

An accurate mathematical solution of the problem, although not impossible, is nevertheless too complicated for general use, and it is not necessary to resort to it, as a very near approximation, fully sufficient for all practical purposes, can be obtained without it.

As the cross-section of the girders, in a bridge of the kind under consideration, is always much greater in proportion to the extensibility and powers of resistance of the material than that of the rods, the middle of the beam may be assumed as a fulcrum, and the strain upon the rods estimated from this point. The uniform weight on the whole bridge being represented by w, the portion at the angle of the tension rod will be $\frac{18}{38} w$, and the strain caused by this weight will be $\frac{18}{38} w$ $(\frac{18 \cdot 25}{3})$ in the direction of the rod. As $w = 110{,}000$ the strain will be 214,133 pounds, requiring (at **10,000** pounds **per square inch)** 21 square inches in the cross-section **of** the **rods,** or in proportion if the strain should **be increased or** diminished.

This is a greater proportion **of iron than is usually** allowed, but it is not too great for **security. The girders cannot oppose** any direct resistance to a cross-strain **without experiencing** flexure; but a railroad bridge should be as rigid as possible, and therefore the rods should be depended upon to resist the whole of the tension, and act as the lower chords of an ordinary bridge. In this way the calculation becomes very simple, and furnishes safe practical **results.**

In the construction of trussed girder bridges, **the** stiffness would be greatly increased by the introduction of diagonal ties **or** braces in the middle rectangular interval, and with this addition, and proportioned upon the principles above illustrated, **it becomes** a **safe,** economical, and in every respect a good)ridge for moderate **spans.**

THE END.

BRIDGE ON PENN^A R. R.
OVER
THE SUSQUEHANNA

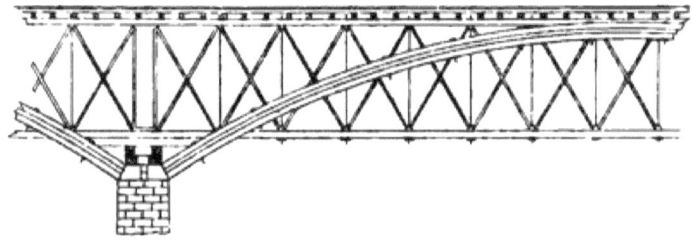

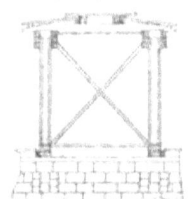

CROSS SECTION

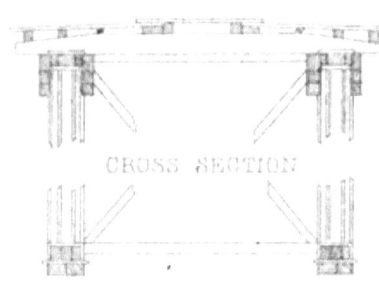

CROSS SECTION

ELEVATION

PLAN OF TOP CHORD.

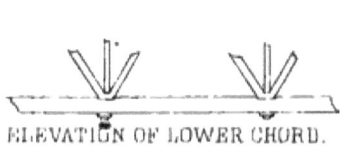

ELEVATION OF LOWER CHORD.

PLAN OF LOWER CHORD.

READING R. R.
ARCHED TRUSS BRIDGE.

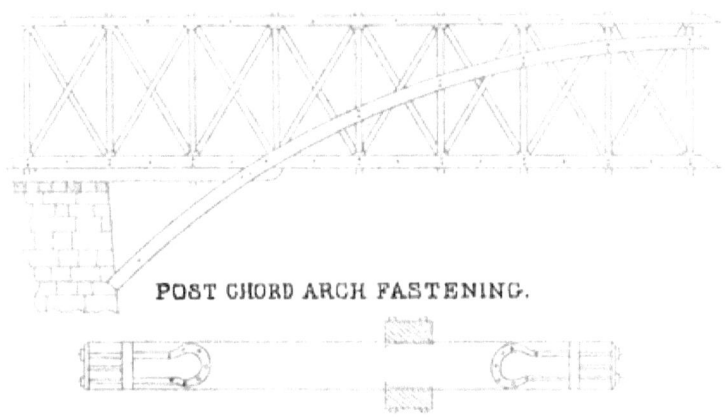

POST CHORD ARCH FASTENING.

COVE RUN VIADUCT

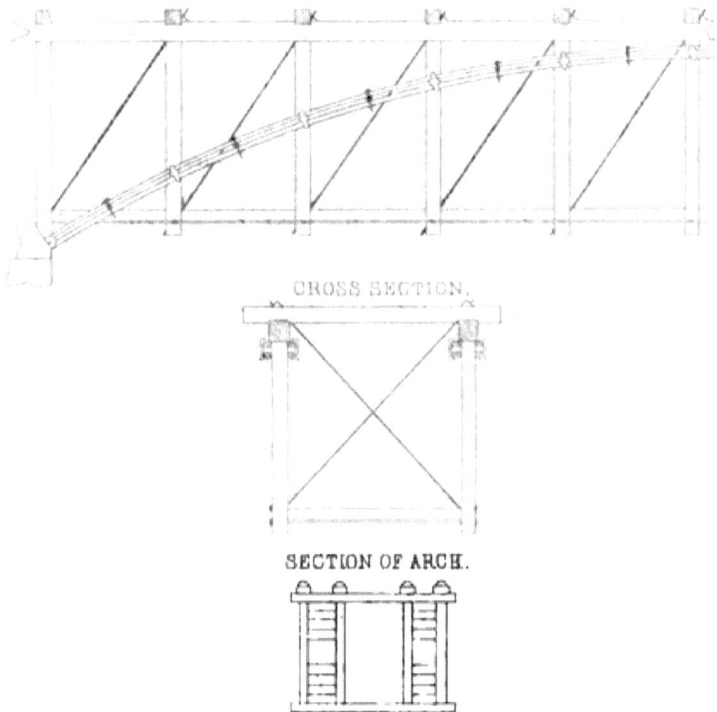

CROSS SECTION.

SECTION OF ARCH.

BALTIMORE & SUSQUEHANNA R.R.
TUBULAR PLATE IRON BRIDGE

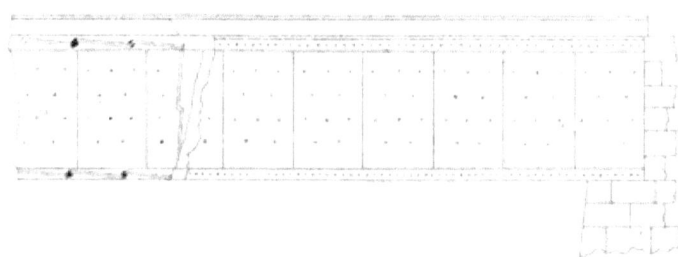

PLAN.

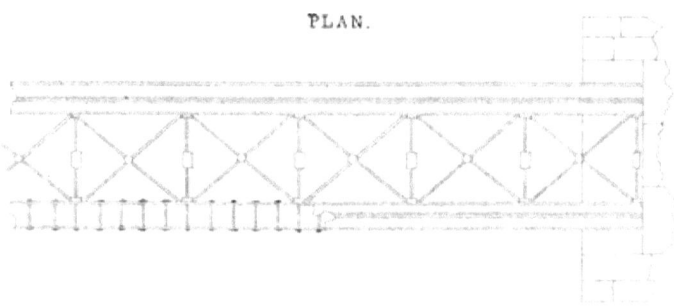

CROSS SECTION

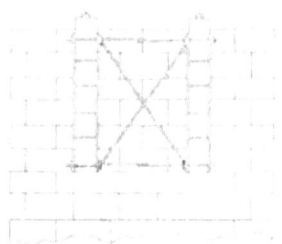

IRON BRIDGE over HARFORD RUN.
Details

TOP CHORD.

Plate 6

PENN^A R. R.
LITTLE JUNIATA BRIDGE.

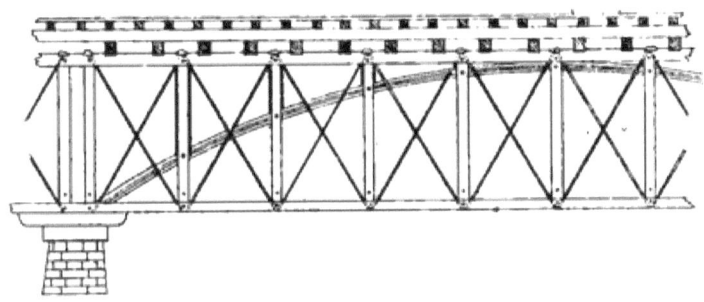

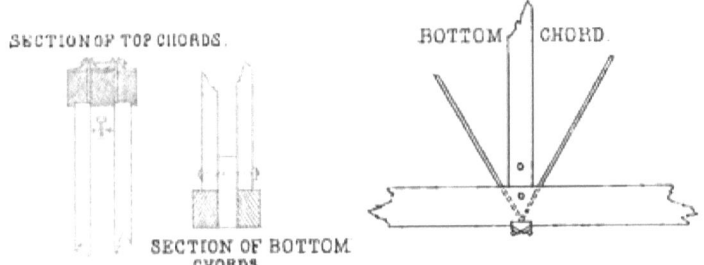

SECTION OF TOP CHORDS.

SECTION OF BOTTOM CHORDS.

BOTTOM CHORD.

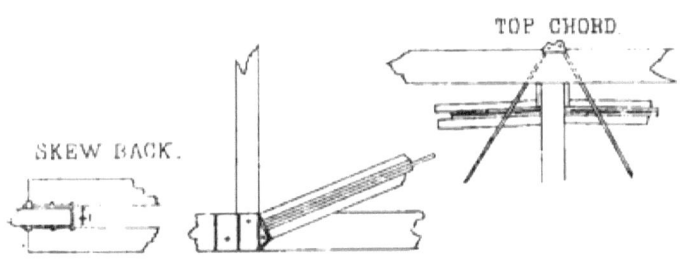

SKEW BACK.

TOP CHORD.

BRIDGE ON PENN[a] R. R.
OVER
SHERMAN'S CREEK

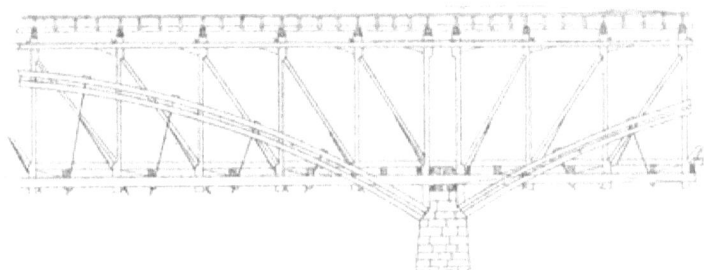

CROSS SECTION.

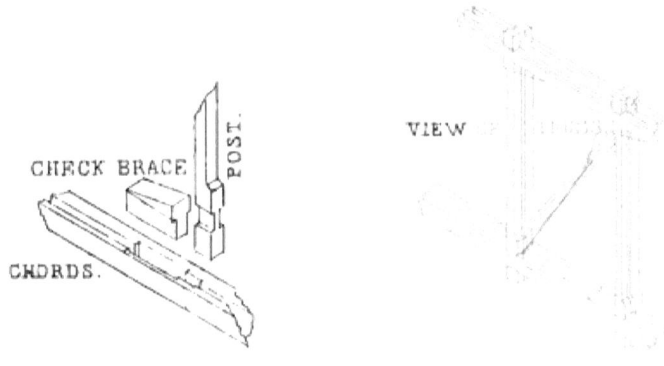

RIDER'S PATENT IRON BRIDGE.

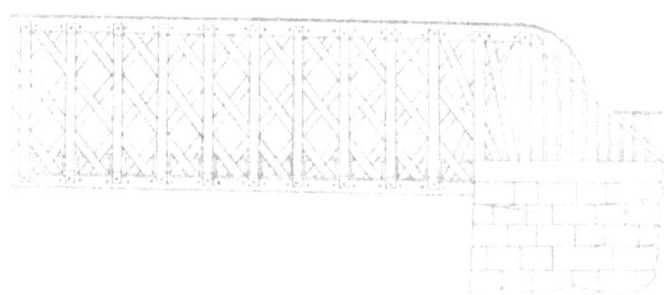

CROSS SECTION

UPPER CHORD

LOWER CHORD

LOWER CHORD

SECTION OF POST

TRENTON BRIDGE.

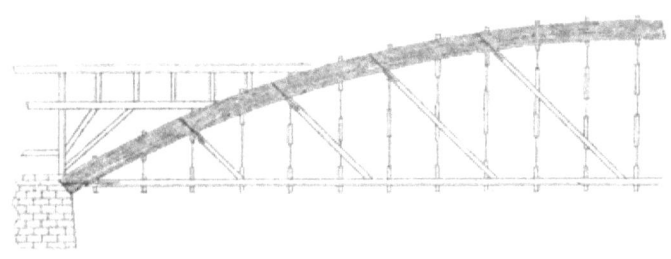

PLAN

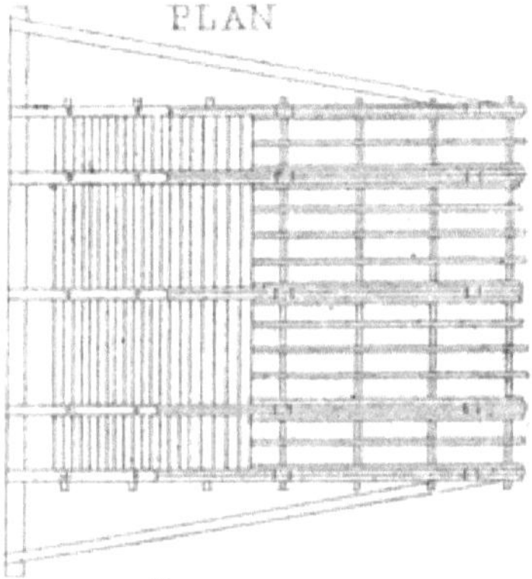

CROSS SECTION

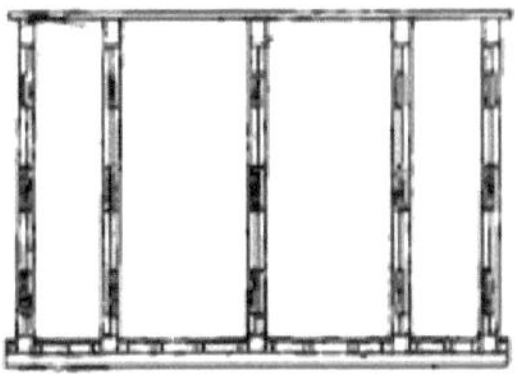

SECTION OF ARCH.

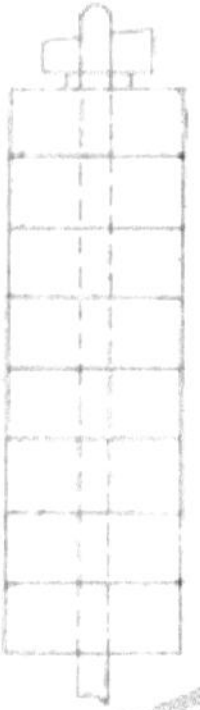

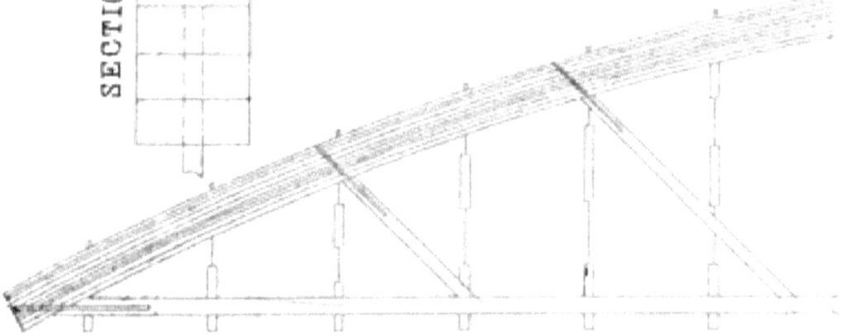

PORTION OF TRUSS

PENNA. R. R.
CANAL BRIDGE.

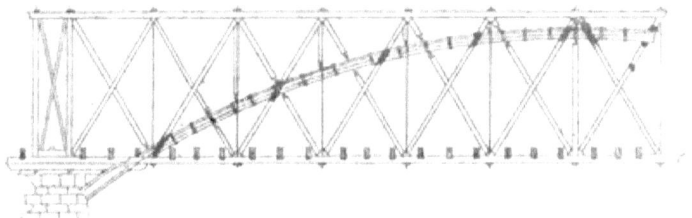

CROSS SECTION.

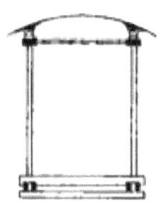

ELEVATION OF ARCH

SECTION

PLAN.

SET SCREW.

SECTION OF LOWER CHORD & ARCH

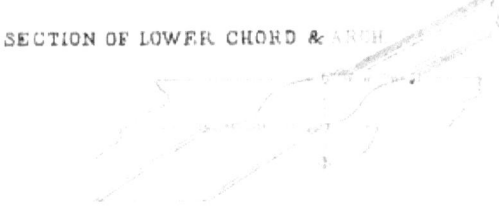

PENNA. R.R.
IRON BRIDGE AT RACOON CREEK.

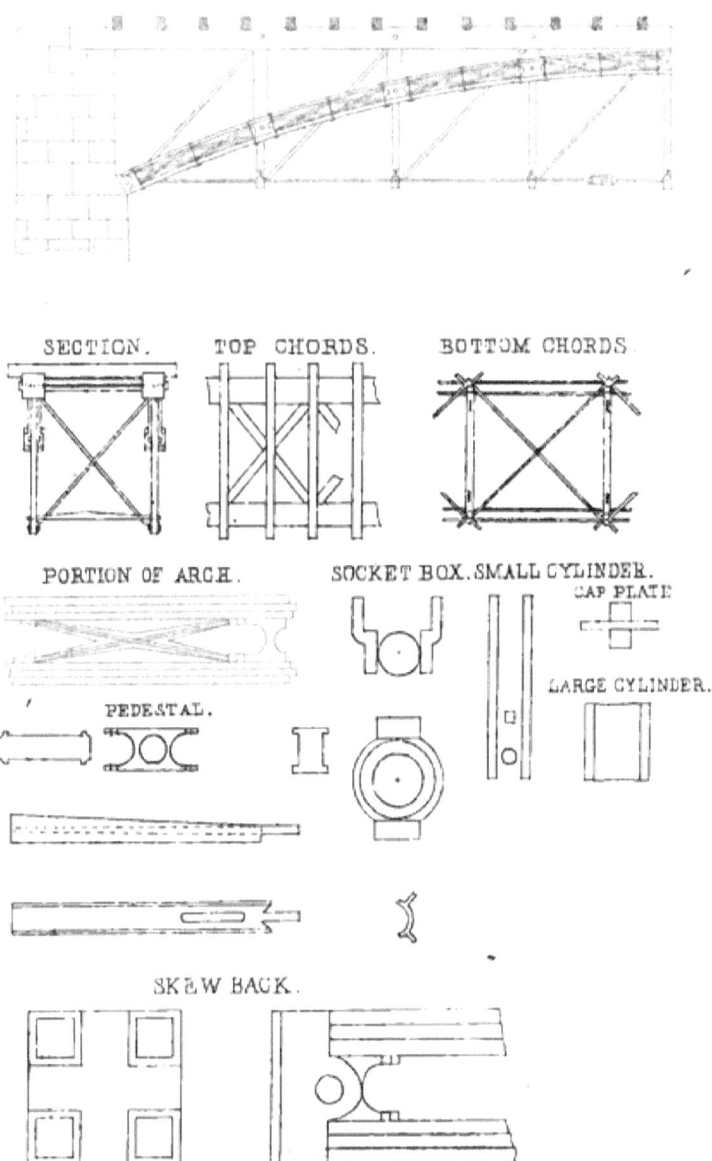

BALTIMORE & OHIO R. R.

IMPROVED ARCH BRACE TRUSS FRAME

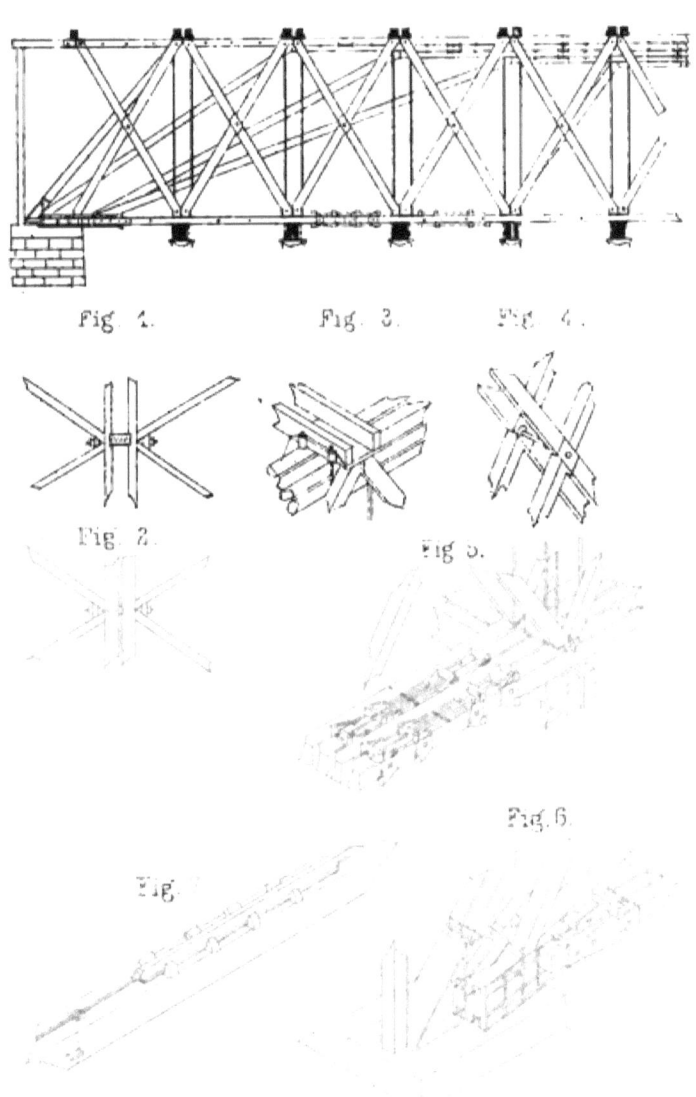

Fig. 1.   Fig. 3.   Fig. 4.

Fig. 2.   Fig. 5.

Fig. 6.

SUSQUEHANNA BRIDGE
AT CLARK'S FERRY.

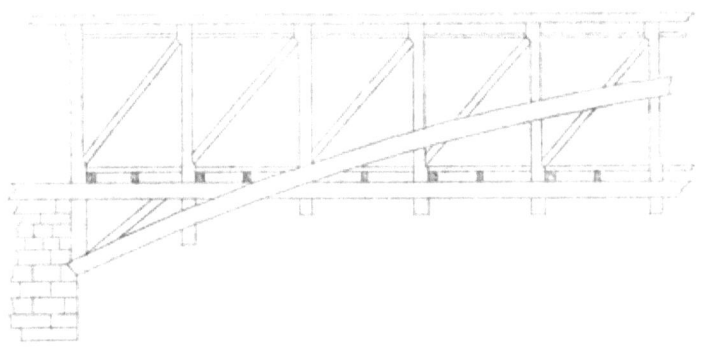

CROSS SECTION.

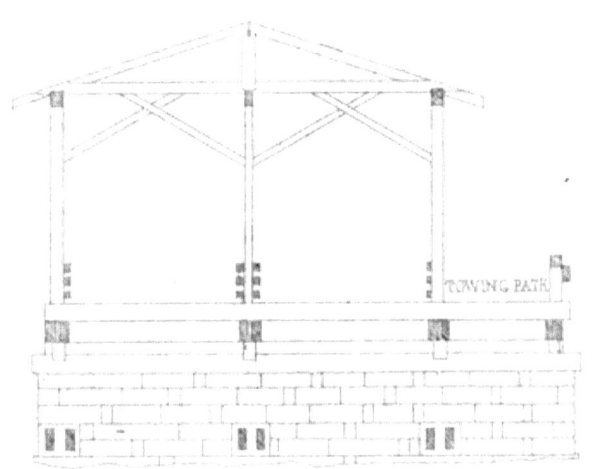

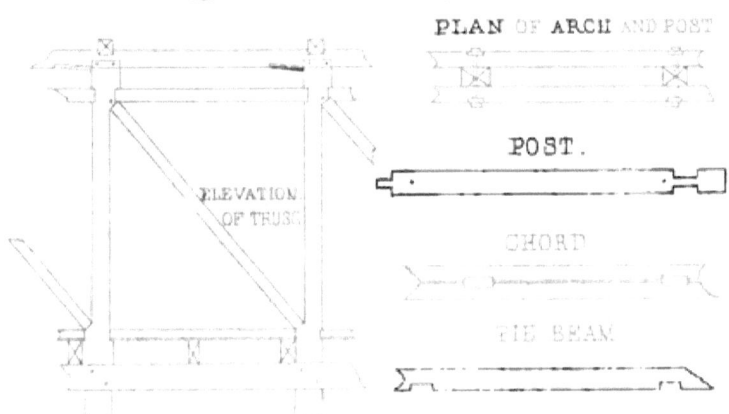

IMPROVED LATTICE TRUSS.

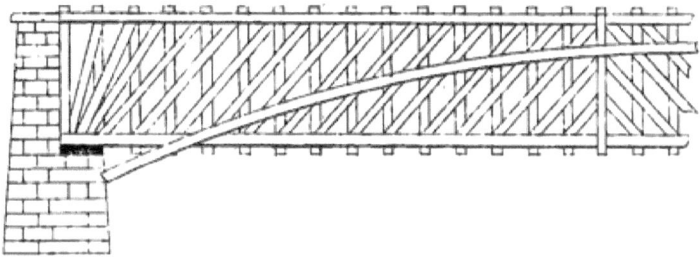

SECTION THROUGH POST.

SECTION THROUGH TIE.

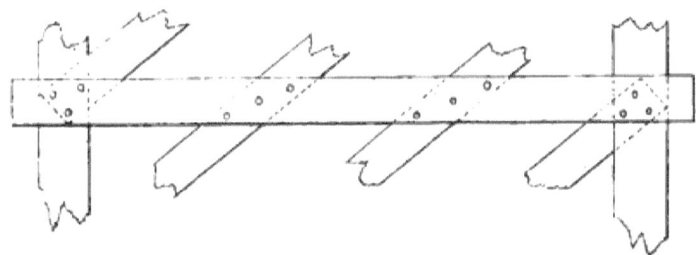

TRUSSED GIRDER BRIDGE.

30 FEET SPAN

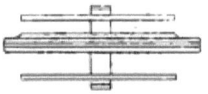

UNDER VIEW OF TENSION RODS &

CROSS SECTION

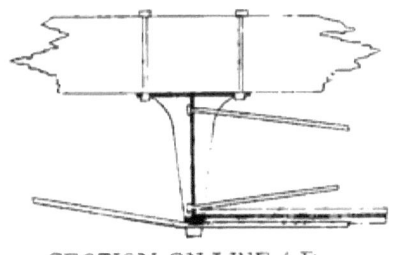

SECTION ON LINE A B

NEW YORK AND ERIE R. R.

GENERAL PLAN FOR BRIDGES

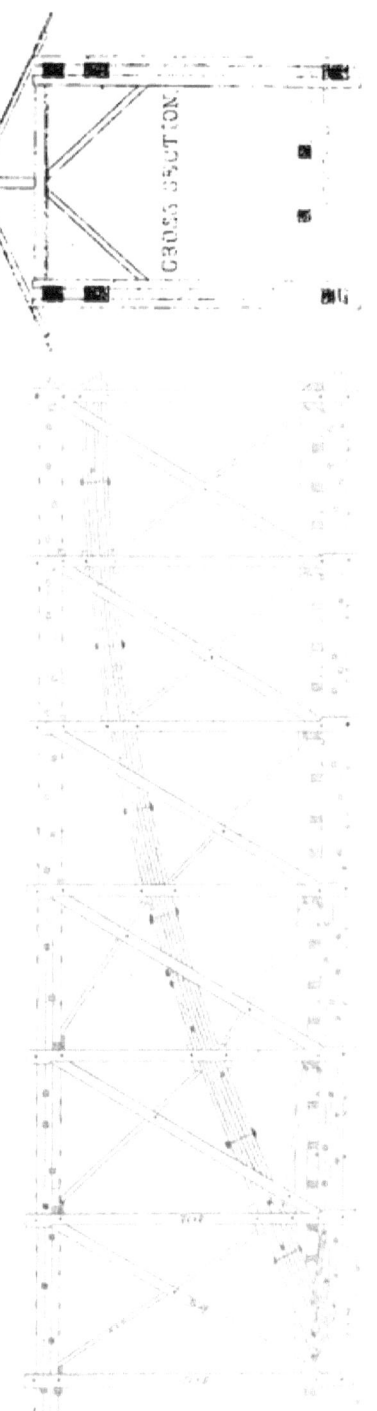

A NEW AND IMPORTANT MOVEMENT IN THE INTEREST OF ACCURATE SCIENCE STUDY.

APPLETONS' SCIENCE TEXT-BOOKS.

D. APPLETON & Co. have the pleasure of announcing that in response to the growing interest in the study of the Natural Sciences, and a demand for improved text-books representing the more accurate phases of scientific knowledge, and the present active and widening field of investigation, they have made arrangements for the publication of a series of text-books to cover the whole field of science-study in High Schools, Academies, and all schools of similar grade.

The author in each separate department has been selected with regard to his especial fitness for the work, and each volume has been prepared with an especial reference to its practical availability for class use and class study in schools. No abridgment of labor or expense has been permitted in the effort to make this series worthy to stand at the head of all educational publications of this kind. Although the various books have been projected with a view to a comprehensive and harmonious series, each volume will be wholly independent of the others, and complete in itself.

The subjects to be comprised are:

PHYSICS,	ANATOMY,
CHEMISTRY,	PHYSIOLOGY
GEOLOGY,	AND HYGIENE,
ZOÖLOGY,	ASTRONOMY,
BOTANY,	MINERALOGY.

NOW READY.

In 12mo. Cloth, $1.25 each.

THE ELEMENTS OF CHEMISTRY. By Professor F. W. CLARKE, Chemist of the United States Geological Survey.

THE ESSENTIALS OF ANATOMY, PHYSIOLOGY, AND HYGIENE. By ROGER S. TRACY, M. D., Health Inspector of the New York Board of Health; author of "Hand-Book of Sanitary Information for House-holders," etc.

To be followed by

A Compend of Geology. By JOSEPH LE CONTE, Professor of Geology and Natural History in the University of California; author of "Elements of Geology," etc.

Elementary Zoology. By C. F. HOLDER, Fellow of the New York Academy of Science, Corresponding Member Linnæan Society, etc.; and J. B. HOLDER, M. D., Curator of Zoölogy of American Museum of Natural History, Central Park, New York.

Other volumes are to follow as rapidly as they can be prepared.

New York: D. APPLETON & CO., 1, 3, & 5 Bond Street.

D. APPLETON & CO.'S PUBLICATIONS.

ROSCOE'S CHEMISTRY—Part II of Volume III.

A Treatise on Chemistry. By H. E. ROSCOE, F. R. S., and C. SCHORLEMMER, F. R. S., Professors of Chemistry in the Victoria University, Owens College, Manchester. Volume III—Part II. THE CHEMISTRY OF THE HYDROCARBONS AND THEIR DERIVATIVES, OR ORGANIC CHEMISTRY. Completing the work. One vol., 8vo, 656 pages. Cloth, $5.00.

⁎ *The previous volumes are:*

Inorganic Chemistry. Vols. I and II. Vol. I. NON-METALLIC ELEMENTS. 8vo. $5.00; Vol. II. Part I. METALS. 8vo. $3.00; Vol. II. Part II. METALS. 8vo. $3.00.

Organic Chemistry. Vol. III. Part I. THE CHEMISTRY OF THE HYDROCARBONS AND THEIR DERIVATIVES, OR ORGANIC CHEMISTRY. 8vo. $5.00.

"It is difficult to praise too highly the selection of materials and their arrangement, or the wealth of illustrations which explain and adorn the text."—*London Academy.*

ELEMENTS OF CHEMISTRY. By Professor F. W. CLARKE, Chemist of the United States Geological Survey. (Appletons' Science Text-Books.) 12mo, cloth, $1.50.

"The author in this text-book presents the difficulties of chemical science to elementary students progressively, and has so arranged the helps in the text and notes that those who have to study without a teacher can readily make certain progress. To those who study the science as a part of their general education, and apply it merely to the every-day applications of life, this book will be found amply complete. To such as seek an advanced course of technical chemical training, this work will serve as a sound, scientific basis for higher study. The experiments cited are simple, and can be readily performed by the student himself with apparatus and materials easily secured. The questions and exercises at the end of the book are not exhaustive, but suggestive and stimulating to further investigation. The book is divided into two parts, Inorganic and Organic Chemistry. An appendix gives a comparative table of English and metric tables, etc."—*Boston Journal of Education.*

TEXT-BOOK OF SYSTEMATIC MINERALOGY. By HILLARY BAUERMAN, F. G. S., Associate of the Royal School of Mines. 16mo, cloth, $2.25.

TEXT-BOOK OF DESCRIPTIVE MINERALOGY. By HILLARY BAUERMAN, F. G. S., Associate of the Royal School of Mines. 16mo, cloth, $2.25.

New York: D. APPLETON & CO., 1, 3, & 5 Bond Street.

D. APPLETON & CO.'S PUBLICATIONS.

A PHYSICAL TREATISE ON ELECTRICITY AND MAGNETISM. By J. E. H. GORDON, B. A. Camb., Member of the International Congress of Electricians, Paris, 1881; Manager of the Electric Light Department of the Telegraph Construction and Maintenance Company. SECOND EDITION, revised, rearranged, and enlarged. Two volumes, 8vo, with about 312 full-page and other Illustrations. Cloth, $10.00.

"There is certainly no book in English—we think there is none in any other language—which covers quite the same ground. It records the most recent advances in the experimental treatment of electrical problems, it describes with minute carefulness the instruments and methods in use in physical laboratories and is prodigal of beautifully executed diagrams and drawings made to scale."—*London Times.*

"The fundamental point in the whole work is its perfect reflection of all that is best in the modern modes of regarding electric and magnetic forces, and in the modern methods of constructing electrical instruments."—*Engineering.*

A PRACTICAL TREATISE ON ELECTRIC LIGHTING. By J. E. H. GORDON, author of "A Physical Treatise on Electricity and Magnetism"; Member of the Paris Congress of Electricians. With Twenty-three full-page Plates, and numerous Illustrations in the Text. 8vo. Cloth, $4.50.

"This work has been in preparation for some two years, and has been modified again and again as the science of which it treats has progressed, in order that it might indicate the state of that science very nearly up to the present date."—*From Preface.*

THE MODERN APPLICATIONS OF ELECTRICITY. By E. HOSPITALIER. New edition, revised, with many Additions. Translated by JULIUS MAIER, Ph. D.

Vol. I. ELECTRIC GENERATORS, ELECTRIC LIGHT.

Vol. II. TELEPHONE: Various Applications, Electrical Transmission of Energy. Two volumes, 8vo. With numerous Illustrations. $8.00.

"M. Hospitalier distinguishes three sources of electricity, namely, the decomposition of metals or other decomposable bodies in acid or alkaline solutions, the transformation of heat into electrical energy, and lastly the conversion of work into current—giving rise to the three specific modes of force styled respectively galvanism, thermo-electricity, and dynamic electricity. He gives a history of the progress of each, from the first crude constructions of the pioneer to the latest and most perfect form of battery, thus furnishing the student of science with a sufficiently copious text-book of the subject, while at the same time affording to the electrical engineer a valuable encyclopædia of his profession. The work presents a most useful and thorough compendium of the principles and practice of electrical engineering, written as only an expert can write, to whom the abstruse by long study has become simple. The translator has acted the part of an editor also, and has added considerable material of value to the original text."—*New York Times.*

New York: D. APPLETON & CO., 1, 3, & 5 Bond Street.

D. APPLETON & CO.'S PUBLICATIONS.

THE ELECTRIC LIGHT: ITS HISTORY, PRODUCTION, AND APPLICATIONS. By ÉM. ALGLAVE and J. BOULARD. Translated from the French by T. O'Conor Sloane. Edited, with Notes and Additions, by C. M. Lungren. With 250 Illustrations. 8vo. Cloth, $5.00.

"Not one of the recent scientific publications was more needed or is more likely to be eagerly welcomed than a clear, exhaustive, and authoritative account of the application of electricity to the production of light. We are indebted to Messrs. Appleton for issuing, in a large volume of 450 pages, illustrated with several hundred woodcuts, an English translation of the well-known treatise by MM. Alglave and Boulard."—*New York Sun.*

ELECTRICITY AND MAGNETISM. By FLEEMING JENKIN, Professor of Engineering in the University of Edinburgh. Illustrated, and Index. With Appendix on the Telephone and Microphone. 12mo. Cloth, $1.50.

"The plan followed in this book is as follows First, a general **synthetical view** of the science **has** been given, in which the main phenomena are described and the terms employed explained. If this portion of the work can be mastered, the student will then be readily able to understand what follows, viz., the description of the apparatus used to measure electrical magnitudes and to produce electricity **under** various conditions. The general theory of electricity is permanent, depending on no hypothesis, and it has been the author's aim to state this general theory in **a** connected manner, and in such simple form that it might be readily understood by practical men."—*From the Introduction.*

ELEMENTARY TREATISE ON NATURAL PHILOSOPHY. By A. PRIVAT DESCHANEL, formerly Professor of Physics in the Lycée Louis-le-Grand, Inspector of the Academy of Paris. Translated and edited, with Extensive Modifications, by J. D. Everett, Professor of **Natural Philosophy in the Queen's** College, Belfast. Sixth edition, **revised, complete in Four Parts.** Illustrated by 783 Engravings on Wood, and Three Colored Plates.

Part I. MECHANICS, HYDROSTATICS, AND PNEUMATICS. Cloth, $1.50.
Part II. HEAT. Cloth, $1.50.
Part III. ELECTRICITY AND MAGNETISM. Cloth, $1.50.
Part IV. SOUND AND LIGHT. Cloth, $1.50.

Complete in one volume, 8vo, with Problems and Index. Cloth, $5.70.

"Systematically arranged, clearly written, **and** admirably illustrated, showing no less than 783 engravings on wood and three colored plates, it forms a model **work** for a class of experimental physics. Far from losing in its English dress **any** of the qualities of matter or style which distinguished it in its original form, it may be said to have gained in the able hands of Professor Everett, both by way of arrangement and of incorporation of fresh matter, without parting in the translation with any of the freshness or force of the author's text."—*Saturday Review.*

New York: D. APPLETON & CO., 1, 3, & 5 Bond Street.

www.ingramcontent.com/pod-product-compliance
Lightning Source LLC
Chambersburg PA
CBHW032046230426
43672CB00009B/1485